South University Library
Richmond Campus
2151 Old Brick Road
Glen Allen, Va 23060

Contemporary Black History

Series editors
Peniel E. Joseph
Department of History
Tufts University
Medford, MA, USA

Yohuru Williams
Department of History
Fairfield University
Fairfield, CT, USA

Maintaining the guiding vision of founding editor Manning Marable, this series features cutting-edge scholarship in Contemporary Black History since 1945, underlining the importance of the study of history as a form of public advocacy and political activism. It focuses on the postwar period, but it also includes international black history, bringing in high-quality scholarship from around the globe. Series books incorporate varied methodologies that lend themselves to narrative richness, such as oral history and ethnography, and include high-quality interdisciplinary scholarship combining disciplines such as African-American Studies, Political Science, Sociology, Ethnic and Women's Studies, Cultural Studies, Anthropology, and Criminal Justice.

More information about this series at
http://www.springer.com/series/14917

Suman Fernando

Institutional Racism in Psychiatry and Clinical Psychology

Race Matters in Mental Health

Suman Fernando
London Metropolitan University
London, UK

Contemporary Black History
ISBN 978-3-319-62727-4   ISBN 978-3-319-62728-1 (eBook)
DOI 10.1007/978-3-319-62728-1

Library of Congress Control Number: 2017947738

© The Editor(s) (if applicable) and The Author(s) 2017
This work is subject to copyright. All rights are solely and exclusively licensed by the Publisher, whether the whole or part of the material is concerned, specifically the rights of translation, reprinting, reuse of illustrations, recitation, broadcasting, reproduction on microfilms or in any other physical way, and transmission or information storage and retrieval, electronic adaptation, computer software, or by similar or dissimilar methodology now known or hereafter developed.
The use of general descriptive names, registered names, trademarks, service marks, etc. in this publication does not imply, even in the absence of a specific statement, that such names are exempt from the relevant protective laws and regulations and therefore free for general use.
The publisher, the authors and the editors are safe to assume that the advice and information in this book are believed to be true and accurate at the date of publication. Neither the publisher nor the authors or the editors give a warranty, express or implied, with respect to the material contained herein or for any errors or omissions that may have been made. The publisher remains neutral with regard to jurisdictional claims in published maps and institutional affiliations.

Cover credit: Canva Pty Ltd/Alamy Stock Photo

Printed on acid-free paper

This Palgrave Macmillan imprint is published by Springer Nature
The registered company is Springer International Publishing AG
The registered company address is: Gewerbestrasse 11, 6330 Cham, Switzerland

To the memory of Darcus Howe, broadcaster and campaigner for the rights of black people in the UK, who died on Saturday 1 April, 2017, just as I was finishing this book.

I never knew Darcus personally, but admired him for his straight talking and grass-roots activism. May he rest in peace.

Preface

This book is somewhat *personal* in that it reflects many of my own experiences and views, both generally about race matters in British society, and about racism (mixed up with cultural bias) in the field of mental health, in particular the practice of psychiatry and clinical psychology.

In the 1960s I embarked on psychiatric training, hoping to build my career as a psychiatrist in the (British) National Health Service (NHS). I decided not to apply for a training post at the so-called centres of excellence in London—I had heard that most prestigious centres followed an old-fashioned institutional approach to patients while some mental hospitals (still very much asylums) were sometimes more forward looking. I opted to work in training posts in mental hospitals in Epsom while attending lectures and seminars at the Institute of Psychiatry. Once I had the diploma in psychological medicine awarded by the Royal Medico-Psychological Association (RMPA), I faced the hurdle of obtaining a senior registrar job, the prelude (after three or four years) to applying for consultant posts in the NHS.

In the 1950s and most of the 1960s, there was a strict racial, and less-strict gender glass ceiling for senior registrar (SR) posts, especially when they were based at teaching hospitals in London. At this point I struck lucky—a Jewish consultant I had worked under looking after inpatients at the base hospital in Epsom and helping with outpatient clinics at the London Jewish Hospital in East London, recommended me to a colleague and friend at the London Hospital. I was appointed SR at the London Hospital, and learned later that this particular consultant, who had clearly spoke up for me at the appointments committee, had himself been the first Jewish doctor to be appointed as a consultant there. This job suited me perfectly because I came to be quite attached to the area that the patients came from, but also because I was able to get funding (through contacts I made) to begin a research study I had planned on socio-cultural aspects of depression among Jewish people in East London. Later, in the 1960s, I worked at Claybury Hospital in Woodford, Essex (where some of the admission wards were organised along the lines of therapeutic community principles); and finally at Chase Farm Hospital (a general hospital) in Enfield, Middlesex.

Many psychiatric institutes in London were operating a race-based glass-ceiling in the 1960s and early 1970s in terms of senior staff appointments; and there were indications that black patients were dissatisfied with the services they received, too. In the late 1970s I became involved in the work of the Transcultural Psychiatry Society (TCPS) in campaigning against racism in mental health services (see 'Transcultural psychiatry in the UK' in Chap. 6), both in terms of the difficulties faced by non-white professionals and by black and brown-skinned users of the mental health services. In the late 1970s, senior posts in psychiatry were gradually opened up to non-white people as legal measures against racial discrimination in employment practices came into effect, but even then black or brown-skinned consultants were a rarity. The race-based glass ceiling at the IOP was not broken until the mid-to-late 1990s, when in the course of a few years, three senior posts were filled by two psychiatrists of Asian origin and one of African-Caribbean origin. Significantly, none of them had participated in the work of the TCPS—it is likely that they would not have attained their posts at the IOP if they *had* done so—but their good fortune may well have had something to do with the work done by the TCPS. In fact one of these psychiatrists told me much later that he thought the work of the TCPS opened doors for him.

In 1986, I was invited to serve on the Mental Health Act Commission (MHAC), a government inspectorate established under progressive new mental health legislation in 1983. There I met and worked with several white people to highlight issues of discrimination and injustice in the mental health services; and I discovered how black professionals who reached higher levels in the mental health field often preferred not to get involved in such activities—mainly because they were afraid for their career prospects. I was fortunate to become chairman of the commission's (then) influential standing committee on race and culture, which devised its first policy on race following episodes of overt racism that had actually occurred within the MHAC itself (see 'Illustrations of institutional racism' in Chap. 6). Unfortunately, when we discovered serious problems in patient care, all that we could do in the MHAC was to report the matter to the managers of the hospital for action to be taken, and then make general recommendations for improvement, or hope for top-down action to be implemented as a result of our including suitable comments in the annual reports that were delivered directly to Parliament. I learned a lot from my service on the MHAC about how psychiatry was experienced by people who were caught up in the system, especially if they were black—something that informed the rest of my time in the discipline. One thing I found difficult to understand was why and how racism has persisted in Western culture for so long; and in particular how it was that psychiatry and clinical psychology, the disciplines that informed the mental health system, a supposedly *medical* system, had come to be as they were—that they were so often experienced as racist by people who use these services. So, while I did the best I could in alleviating the oppression that people using the mental health services felt (which was not much because for much of the time I had to adhere to what I felt was a flawed system), I read about its origins and background, integrating what I found out with my observations and personal experiences of the system itself. This book is an attempt to share my thoughts in the form of a historical account.

London, UK
Suman Fernando

Acknowledgements

I am indebted to works of scholarship in the fields of history, psychology, psychiatry, religion and sociology, especially those works that take a critical approach to psychology and psychiatry and indeed to race and culture; and those that cut across disciplines. In particular I would like to express my appreciation of some classics in the field of race and culture, namely the works of Frantz Fanon, Stuart Hall, Edward Said and Homi Bhabha; and the book that first led me to examine the background to racism in psychiatry—Alexander Thomas and Samuel Sillen's *Racism and Psychiatry, the Black Patient—Separate and Unequal*, published in 1972.

I have included in my book a few personal accounts and true stories in order to illustrate the ways in which the personal is political and vice versa; and I wish to acknowledge the support and advice I received from colleagues and friends during some of the incidents that I describe. All in all, I have had the support throughout my writing career of numerous colleagues and friends, especially people I worked with in the Transcultural Psychiatry Society (UK) for over twenty years until it was disbanded by mutual agreement in 2008. And I am grateful to many colleagues in the field of mental health for their encouragement whilst writing this book—in particular I wish to mention the support of Roy Moodley and Martha Ocampo of Toronto. Finally, I am grateful to Megan Laddusaw, Commissioning Editor and Christine Pardue, Editorial Assistant at Palgrave Macmillan for their patience, advice and help in taking this book to completion, and to my wife Frances and daughter Siri for their support and encouragement during its writing.

Contents

1 Introduction   1
   References   9

2 How 'Race' Began, and the Emergence of Psychiatry and Clinical Psychology   11
   2.1   *Race Thinking*   12
   2.2   *Exploration, Colonialism, Race-Slavery*   12
   2.3   *The European Enlightenment*   17
   2.4   *Scientific Racism*   19
   2.5   *Origins of Western Psychology and Psychiatry*   20
   2.6   *The Scientific Paradigm*   23
   2.7   *Biologisation of Mind*   26
   2.8   *Sociopolitical Context*   27
   2.9   *Limitations of Knowledge*   30
   2.10   *Modern Psychiatry and Clinical Psychology*   33
   References   34

3 Race Thinking and Racism Become the Norm   39
   3.1   *Effects of Colonisation*   39
   3.2   *Power of Racism*   41
   3.3   *Distortions of History*   43
   3.4   *The Arts and Nineteenth-Century Sociology*   45
   3.5   *Nineteenth-Century Psychology and Psychiatry*   46
   3.6   *Inherited Instincts*   48

	3.7	Race Psychology	49
	3.8	Mental Pathology and the Construction of Race-Linked Illnesses	50
	References		54
4	**New Racisms Appear in the 1960s**	59
	4.1	Transformations After WWII	59
	4.2	American Social Studies	65
	4.3	Black Protest in the UK	69
	4.4	Definitions of Racism and Race	71
	4.5	New Racisms in the UK	74
	4.6	Racist IQ Movement	76
	4.7	Alleged Mentality of Black People	78
	4.8	Racism in Cultural Research	82
	4.9	Conclusions	84
	References		86
5	**Racism in a Context of Multiculturalism**	91
	5.1	Discrimination, Diagnosis and Power	91
	5.2	Ethnic Issues in Mental Health Services	93
	5.3	Racialisation	95
	5.4	Racism in 'Psy' Research	98
	5.5	Manipulation of Research Findings	100
	5.6	Explanations for 'Schizophrenia' in Black People	103
	5.7	Racialisation of the Schizophrenia Diagnosis	105
	References		106
6	**Struggle Against Racism in the UK**	111
	6.1	The Macpherson Report	111
	6.2	Transcultural Psychiatry in the UK	114
	6.3	Action on Apartheid	117
	6.4	Action by Black Professionals	117
	6.5	The Black Voluntary Sector (BVS)	119
	6.6	Institutional Action	121
	6.7	Government Action	122
	6.8	Illustrations of Institutional Racism	124
		6.8.1 The MOST Project	124
		6.8.2 The Ipamo Project	126

		6.8.3 Racist Exploitation of a Black Organisation	127
		6.8.4 Racism in a Government Body	128
	6.9	Race Matters in Professional Associations	130
	References		132

7 Persistence of Racism Through White Power — 135
7.1 Controlling Racialised Minorities — 136
7.2 Employment in the Mental Health System — 138
7.3 Institutional Racism in the Department of Health (DOH) — 140
7.4 Black People in White-Dominated Systems — 143
7.5 How Whiteness Operates — 145
7.6 Privilege and Power — 146
7.7 White Knowledge — 148
References — 150

8 Racism Post-9/11 — 153
8.1 Diasporic Identities, Nationalisms and Multiculturalism — 154
8.2 Obama Years — 156
8.3 Rise of the Political Right — 158
8.4 Civil Unrest — 161
8.5 Changes in the Field of Mental Health — 161
8.6 Racist Conclusions of Psychiatric Research — 162
8.7 Racism of a Psychology Report — 163
8.8 Islamophobia — 167
8.9 The 'Psy' Disciplines and Islamophobia — 169
8.10 Conclusions — 173
References — 173

9 Racism with the Advent of Trump and After Brexit — 181
9.1 New Era of Unashamed Racism? — 182
9.2 Why Racism Has Persisted — 184
9.3 Future of the 'Psy' Disciplines — 185
9.4 Conclusions — 186
References — 189

Bibliography	193
Author Index	197
Subject Index	203

List of Figures

Fig. 2.1　Historic context of psychiatry and psychology　　　14

List of Tables

Table 2.1	Scientific paradigm	24
Table 4.1	Theories of black racial inferiority	66
Table 5.1	Racial inequalities in the UK	94
Table 7.1	Ethnicity and diagnosis	137
Table 7.2	Ethnicity and 'stop and search'	137
Table 7.3	British citizens in prison in England and Wales	137
Table 7.4	School exclusions	137

CHAPTER 1

Introduction

The notion of 'race' has been problematic for a long time and it has bred racism—an undoubted reality for many people from (what these days are called) racialised groups (see 'Racialisation' in Chap. 5). Today, racism is like a sin, everyone seems to be against it—or at least says they are. Yet only fairly recently—even into the 1960s—some people who saw themselves as respectable were not ashamed to be racist. As a newly-wed couple in the early 1960s, my wife and I presented at a hotel in the town of Teignmouth in the English county of Devon and asked for a room at an expensive hotel. The landlord politely informed me that he operated a 'colour bar' because (according to him) that was what his clients wanted. That was in the early 1960s, before the 1965 Race Relations Act (Hepple 1966) that outlawed racial discrimination—the law was strengthened later, the Race Relations Act (1976) making it unlawful to refuse hospitality at a hotel on the grounds of race. The 1965 Act set a standard of behaviour that helped promote the changes in public attitudes that followed over the next two decades. It is well to remember that the 1960s was the time of legalised race segregation in the USA—it was the Jim Crow era (see 'Exploration, colonialism, race-slavery' in Chap. 2), so the overt racism in UK at the time was relatively mild in comparison to that in the USA—but yet it was *unashamed* and in one's face. In late 1960s into the 1970s, a feeling of shame about being racist gradually infiltrated public behaviour in the UK, although it was often merely about being *seen* to be racist, and many people became careful in the language they used (later called 'political correctness' of language—see Weigel 2016).

© The Author(s) 2017
S. Fernando, *Institutional Racism in Psychiatry and Clinical Psychology*, Contemporary Black History,
DOI 10.1007/978-3-319-62728-1_1

In the 1960s, I became interested in researching cultural issues in mental health field. Since I was working at the London Jewish Hospital I was persuaded by a colleague to focus on (and interested in) Jewish people and fortunate to be supported by the new professor of psychiatry at the London Hospital in organising a research project into cultural aspects of 'depression' that studied people who had been diagnosed as depressed. Essentially, I planned to compare Jewish and non-Jewish individuals in terms of nominal religious practices and ways of living. The project took me about four years to complete and was successfully presented for a MD degree much later. The subjects of research were Jews and non-Jews who had been born in a defined part of East London and were from the same social-class background—in other words the groups were controlled for social background. However, as I talked with the people concerned (the subjects of the research) it seemed to me that the experience of the Jews I interviewed had been very different to that of the non-Jews. The former had experienced the dreadful anti-Semitism that was prevalent in East London in the first half of the twentieth century. In fact the background life experiences of the two groups (Jews and non-Jews) were very different—so in this respect they were *not* controlled for social background. I realised then that studying cultural differences cannot ignore racism in people's experiences—'race' and 'culture' always interact and must be taken together. This lesson stayed with me ever since and has affected much of my later research and writing.

Racism reappears again and again in various manifestations in the UK, if not the world over. And it is the UK, and to a lesser extent the USA, that I focus on in this book, looking in particular at mental health services and the two disciplines that underpin these services—clinical psychology and psychiatry (which I sometimes refer to as the 'psy' disciplines). I have been able to approach the writing of this book from *within* (as it were) the field of mental health because I was a practising psychiatrist for over twenty years in a multicultural area of London, and was drawn into exploring the ways in which racism was manifested in clinical work, including my own, and, more seriously, the way that mental health services were experienced by black and minority ethnic (BME) people (the acronymic term BME is applied to people seen as the racial 'Other'; *racialised* people, black people in the language of 1960s UK (see 'Racialisation' in Chap. 5).

Although the notion of 'race' as a reliable attribute of human beings is problematic (hence the inverted commas round the word in this instance), it seems to persist as a sociopolitical phenomenon (Omi and Winant 1994, 2015) and it is commonly accepted as a social reality (see 'Definitions of racism and race' in Chap. 4). Racism used to be specifically about discrimination, implying the inferiority or even oppression of people designated as belonging to a 'race' apart from the 'white race'—and hence racism is always associated with the notion of white supremacy (see 'Privilege and power' in Chap. 7)—but today (2017) it refers mostly to a variety of behaviours and attitudes expressed often by gesture, innuendo or just by the way a person *looks*; it does not necessarily refer to overt racist behaviour such as physical attacks, spitting or hurling racist epithets, although of course these would be definitely racist and are unfortunately becoming much commoner now in the UK than they were a few years ago, in my experience. The power and significance of the look is amplified by Fanon (1952) in *Black Skin, White Masks*.

My academic and professional background is in the field of mental health and my interests have long been centred on the way the clinical disciplines of psychology and psychiatry have developed in Western culture, especially in their cultural and social roots and dimensions; and on the history of mental health, especially the services for providing help and support for people who are deemed to suffer from (what are commonly called) mental health problems. Throughout most of my professional life, nearly all of it spent in the UK, I have been mindful, sometimes intensely so, of the interplay between individual personal experiences, professional responsibilities, the lives of patients and clients (of the mental health services) and the sociopolitical forces and activities that connect with mental health work. In other words, the personal, the political, the academic and professional have meshed together in much of my adult life. And in writing, I find it difficult to separate them out into silos. So my approach to the writing of this book is (a) to look at the *history* of 'race' and racism, and how they impact on the clinical practice of psychology and psychiatry (the 'psy' disciplines); and (b) to explore the issues involved by presenting a mixture of historical facts researched with as much care as I can muster, from a number of sources; the lessons I learned from clinical experiences and from talking with people I have met as users of mental health services; and mindful observations of real life in the politics of the British mental health system—all told against the background of a wider social and political context (mainly that in the UK) and told, as much as possible, *like it is*.

The rest of the book is set out vaguely in terms of a sequence of events and happenings starting from the 'beginning' (another problematic notion) and building up to the present, as one would expect a history book should be. Bearing this in mind, my approach (stated above) means that there will be toing and froing across time, but hopefully with adequate cross-referencing. The book traces in broad outline how racism grew in importance, then went underground (as it were) by becoming reconstructed in ever newer forms, interacting all the time with the anti-racist forces in society; while the original biological understanding of the notion of race changed, to become socially and politically constructed.

Racism was one of the powerful forces that drove the slave ships on the prolonged Atlantic slave trade, and almost concurrently (and for even longer) underpinned the massive project of colonial conquest, exploitation and often appropriation, by white Europeans, of lands and property (including the intellectual property) of people seen (by whites) as *not* being 'white'—almost as if they (the whites) felt these 'other' things and 'other' people belonged to white people; even the history of humankind came to be seen (by white people) as belonging to them. Something often forgotten is that racist ideas were heard loud and clear at the dawn of *modern* European culture at the time of its so-called 'Enlightenment', and was (and still is) deeply embedded in what were—and possibly still are to a great extent—'European values'. Certainly racism was one of the values that European nations exhibited right up to the mid-twentieth century, but since then (roughly since the end of the Second World War, or WWII) European culture has steadily moved away from its racist inheritance, albeit with a revival during the past decade which, hopefully is a temporary setback. Cultural shifts that occurred in Europe during the Enlightenment gave rise to (among others) the professional fields of psychiatry and clinical psychology—a duo that developed in tandem—both fields being akin to medicine in their concern with the observation and treatment, often referred to within the clinical professions as 'management', of people called 'patients' or 'clients'.

Although after WWII and the fall of European colonial empires, racism was challenged in many walks of life in Western countries, racism in ever newer forms continues to impact on many aspects of Western society (including the activities of the 'psy' disciplines), having become largely institutional and subtle in its expression and manifestation (see 'Definitions of racism and race' in Chap. 4). Also, more recently, racism has extended to affect groups of people apart from those traditionally seen in European societies as the racial 'Other', such as immigrants, refugees and people seen as 'Muslims', Romany communities and so on. To a greater or lesser extent these groups are racialised (see 'Racialisation' in Chap. 5). And, during the past decade—perhaps since the attacks on the twin towers in New York (9/11) and the so-called war against terrorism—there has been a rise in

Euro-America of less subtle forms of racism, racism in the form of (for example) racist attacks supposed to punish terrorists, or terrorist sympathisers seen as 'radicals', or just people seen as representing them in some way (the notion being sometimes fanned for political reasons) by being immigrants—hence attacks on immigrants. Along with this, there has recently (2001 onwards) been a greater tolerance in official circles, at middle-class dinner tables and in professional institutions, of institutional racism including subtle allegorical references to 'race'. All this, while an impression has been created by the powers to be that the Western world is no longer racist—that it is 'post-race', represented by the fact that Americans elected a black man as president in 2008, and then, to the surprise of much of the world, a racist white supremacist eight years later (see 'Obama years' in Chap. 8 and 'New era of unashamed racism?' in Chap. 9). The rest of this book covers many of these matters briefly, but I hope succinctly, and attempts to see their connection to the ups and downs of racism in the 'psy' disciplines.

Chapter 2, 'How "Race" Began, and the Emergence of Psychiatry and Clinical Psychology', traces the development (a) of the notion of 'race' through to its firm embodiment in racism via 'race thinking'; and (b) of the disciplines of clinical psychology and psychiatry through the study of 'madness', in a context of the confinement of various types of people who were socially excluded and forced into asylums. The chapter touches on how racism was rooted in the Atlantic slave trade, and vice versa; how slavery provided the wealth for the economic power that generated colonialism; and how racism became diversified and stabilised through the near-total domination of the world by European power. It describes in brief how the values attributed to 'reason' and 'unreason' in the way people (in the West) lived their lives, changed at around the time of the Enlightenment; how beliefs and behaviours that were deemed unacceptable and alien to polite (white-European) society became symptoms and pathologies in models of illness or formulations of deviance concocted by doctors ('alienists' and 'mad-doctors', later called psychiatrists) and clinical psychologists; how sociopolitical-cultural structures came into being in the West to become (what we now call) clinical psychology and psychiatry; and finally how the biomedical ways of interpreting the lived experience of human beings has become embedded in Western culture—together with, and bound up with, racism.

Chapter 3, 'Race Thinking and Racism Become the Norm', begins by describing the nature and power of the colonial projects around the world and how they resulted in racism and white supremacy becoming political and social norms that would dominate social and cultural development—changes that were then spread around the world. This chapter discusses how the achievements of African, Asian and pre-Columbian cultures were discredited, and the very writing of history was distorted to suit white supremacist thinking. Anything to do with progress and civilisation became, by definition, Eurocentrist (Euro-culture now including white America). The arts, literature and the budding discipline of sociology were all infiltrated by racist notions and white supremacist thinking. As the European Enlightenment came on stream, racism became a prominent value voiced by leading (European) thinkers and writers, carried over to the colonised white-European USA. Inevitably, the new disciplines that were formed, including psychology and psychiatry, were permeated by racism—something that the chapter illustrates with several examples, such as the theory of inherited instincts, race psychology in the USA and contemporary explanations of the genesis of various psychological and psychiatric illnesses.

Chapter 4, 'New Racisms Appear in the 1960s', is concerned largely with the changing scene following the end of WWII—a political and social (and possibly cultural) watershed in European civilisation. These changes can be seen in the demographic transformations in Europe that resulted when people who were regarded as the racial 'Other' came to live in the West, and in the civil rights movement in the USA, which brought movement towards the liberation of the formerly enslaved, and later oppressed and disempowered, African Americans—both events taking place in a context of persisting racism that was changing in style and expression. These changes were matched by important political events such as the creation of the United Nations; the acceptance by ex-colonial (white) nations that human rights, irrespective of racial origin or appearance, needed to be given importance (something absent in their thinking before WWII); and the gradual but inconsistent economic rise of some previously impoverished communities in the former colonial world (the 'Third World'). The chapter discusses the changes and struggles against racism in the field of mental health in the new context of this rapidly changing world. It covers some aspects of American sociological studies that appeared liberal in outlook but revealed hidden racism; black protest movements in the mental health field in the UK; and the revival of the racist IQ movement under a camouflage of liberal ideas—all of which showed up the nature of the new racisms that were developing as overt racism became unpopular and something to be ashamed of, and was covered up by language termed 'political correctness'.

Chapter 5 ('Racism in a Context of Multiculturalism') is mainly about the politics of race and culture, and refers largely to the UK, where multiculturalism was official government policy in the 1980s and 1990s, a tool with which to improve 'race' relations' and deal with ethnic tensions that were evident, not least in the field of mental health. In this chapter I discuss the interplay between discrimination, diagnosis and power in the field of mental health using examples from South Africa during apartheid and the former Soviet Union. Then I discuss—briefly, because it is a large topic about which I have written elsewhere—the ethnic inequalities in the UK's mental health services, focusing particularly on the over-representation of black people in the psychiatric category of schizophrenia, now virtually seen as a 'black disease' not just in the UK but also in parts of the USA and probably on the continent of Europe—that is, wherever institutional racism with associated white supremacist ideologies seems to be active. The chapter argues that racism in the 'psy' disciplines is one of the main reasons why modern psychiatry is unsatisfactory and why 'talking treatments' and other psychological therapies in clinical psychology appear to be ineffective and poorly regarded by people from non-Western cultural backgrounds (Asian and African diasporas) living in Western countries. The chapter discusses the process of racialisation, whereby various groups of people are treated as if they are 'races', suffering varying degrees of oppression and discrimination. I outline the complicated, and sometimes subtle, ways in which racism is manifested in psychological and psychiatric research; and I quote illustrations of how research findings are sometimes manipulated to suit the prejudices of the people carrying them out. Finally, the concept and diagnosis of schizophrenia is considered in some detail with examples of how it has been both racialised and politicised.

Chapter 6 ('Struggle against Racism in the UK') begins by describing a landmark report on institutional racism in the police force in London and then goes on to consider attempts to grapple with racism by people and organisations in the UK, illustrated by personal experiences. These include the work of the Transcultural Psychiatry Society (UK) and that of both black and white professionals committed to anti-racism. The influence of voluntary organisations run by black and other minority-ethnic people (the black voluntary sector) during the 1980s and 1990s is outlined. There are discussions on work done within the Royal College of Psychiatrists and through the work of a government inspectorate, the Mental Health Act Commission (MHAC); and the chapter considers the working of a largely ineffective government strategy, 'Delivering Race Equality', that was operational from 2003 to 2007. Institutional racism is illustrated by four accounts: two projects set up and staffed mainly by black people; one about the exploitation of a black organisation by a (white) university; and one that describes the struggles within a government inspectorate (the MHAC) during the 1990s. The chapter ends by discussing the way race matters are handled in general by the professional bodies associated with clinical psychology and psychiatry.

Chapter 7 ('Persistence of Racism through White Power') starts by citing how badly the Western world has done in trying to live up to the expectations of the UN (1965) Convention on the Elimination of All Kinds of Racial Discrimination. The chapter lists tables that outline statistics of racial inequalities in the systems of mental health, policing, criminal justice and education in the UK; when these statistics are added up, they argue for all four being geared towards socially controlling black people—racialised minorities. The chapter then discusses, with illustrations from personal experience, how racism is rife in employment in the health services, the department of health and white-dominated systems in general. It goes on to discuss how black people can influence the policy and activities of white-dominated institutions—and the pitfalls of trying to do so. Discussion of how whiteness operates indicates how white supremacy became embedded in Western culture and is now reflected in most, if not all, systems in the UK and North America. Sections on privilege and power and white knowledge complete the chapter.

Chapter 8 ('Racism Post-9/11') deals with some of the happenings since the West launched (what George W. Bush, then President of the USA called) a 'crusade', evoking 'a "clash of civilizations" between Christians and Muslims' (Ford 2001). But then came the Obama years, carrying hope of progress towards a non-racist Euro-America—a hope that was unfulfilled. The chapter discusses the rise of the political right, which exploded with the UK's decision to leave the European Union (Brexit) and the election of Trump as President of the USA. Racism is once again being openly expressed, as it was prior to the mid-twentieth century. The promise of change following the end of racist colonialism seems not to have been fulfilled. The chapter contains sections on civil unrest in the UK and the apparent worsening of racism in the clinical disciplines and in mental health services. I discuss two highly significant events in which I have been involved personally—one concerning a psychiatric study and one a clinical psychology report—which are described in some detail in order to illustrate the fact that institutionally racist attitudes in the clinical fields of psychology and psychiatry, that may have been hidden by the use of politically correct language, are now emerging into the open once again in the 'psy' disciplines. After discussing Islamophobia, I discuss how psychiatry and clinical psychology may be being drawn into the social control of people thought to have been radicalised into having extremist views—and the danger this shift presents to professional practice and to society at large. Finally I conclude the chapter with a summary of how and why the racist tradition in the West and racism in the 'psy' disciplines continue to the present day.

Chapter 9 ('Racism with the Advent of Trump and After Brexit) was written during the politically fraught, socially unstable spring of 2017. In this chapter I wonder whether the trajectory of a definite, although admittedly slow, movement away from racism, has abruptly shifted; and whether we may be heading towards a new era of unashamed racism—and if so why this change is happening, and how it is that racism has persisted for so long in Western culture. I can see the struggle against racism continuing indefinitely and I wonder whether it is feasible to think of eliminating it from psychiatry and clinical psychology until we can bring about paradigmatic changes in the structures of the disciplines themselves. I speculate on the future of the 'psy' disciplines themselves, in the light of new knowledge in the neurosciences—and how racism may play out if these disciplines change fundamentally in their structures. The professional bodies concerned with psychiatry and clinical psychology seem institutionally unable to see racism as a challenge; with this in mind, I present some views on how the struggle against racism in psychiatry and clinical psychology may be pursued in the future. The way forward may be to confront, primarily, issues of power, rather than just racism alone; and to focus for the time being on universalising, decolonising, the knowledge base of the disciplines of psychiatry and clinical psychology. It is possible that racism cannot be eliminated from the 'psy' disciplines until those disciplines themselves undergo paradigmatic change, so the anti-racism struggle should join forces with efforts to bring about changes in psychiatry and clinical psychology.

REFERENCES

Fanon, F. (1952). *Peau noire, masques blancs*, (Editions de Seuil, Paris. C. L. Markmann Trans., 1967), *Black skin, white masks*. New York: Grove Press.

Ford, P. (2001). Europe cringes at Bush "crusade" against terrorists. *The Christian Science Monitor*, September 19, 2001. Retrieved on March 6, 2017 from http://www.csmonitor.com/2001/0919/p12s2-woeu.html.

Hepple, B. A. (1966). Race Relations Act 1965. *The Modern Law Review, 29*(3), 306–314.
Omi, M., & Winant, H. (1994). *Racial Formation in the United States* (2nd ed.). London and New York: Routledge.
Omi, M., & Winant, H. (2015). *Racial Formation in the United State* (3rd ed.). New York and London: Routledge.
Race Relations Act. (1976). London: Her Majesty's Stationery Office.
UN (United Nations). (1965). *United Nations International Convention on the Elimination of All Forms of Racial Discrimination (CERD)* Geneva Office of the High Commissioner, United Nations Human Rights. Retrieved on February 10, 2017 from http://www.ohchr.org/EN/ProfessionalInterest/Pages/CERD.aspx.
Weigel, M. (2016). 'Political correctness: how the right invented a phantom enemy' The Guardian, 30 November 2016. Retrieved on December 10, 2016 from https://www.theguardian.com/us-news/2016/nov/30/political-correctness-how-the-right-invented-phantom-enemy-donald-trump.

CHAPTER 2

How 'Race' Began, and the Emergence of Psychiatry and Clinical Psychology

The etymological roots of the word 'race' can be traced to the Spanish *raza*, used in Spain in the Middle Ages 'to refer to different breeds of dogs, horses, and, when referring to human populations, Moors and Jews' (Gordon 2015, p. 136), a vague racial awareness (in its modern sense) having developed in Europe during the Middle Ages in connection with these two groups of people as the non-Christian 'Other' (Banton 1987). Ryszard Kapuściński (2008), Polish-born journalist and traveller, has examined the idea of the 'Other', derived (in Western culture) during contact between the West and the Rest. Stuart Hall (1996), British sociologist, argues that in a bipolar discourse of dividing the world, 'emergence of an idea of "the West" [vs the Rest] is a *historical*, not a geographical, construct ... central to the Enlightenment' that emphasised European uniqueness and superiority compared to anything non-Western (pp. 186–187, emphasis in original). Kapuściński (2008) traces the contact between the two worlds of the West and the Rest in terms of eras (periods of history). The first era, which lasted roughly until the fifteenth century, saw such contact mainly on trade routes or diplomatic missions; the second era, during European exploration, was the 'period of conquest, slaughter and plunder, the real dark ages in relations between Europeans and Others' (pp. 26–27). The latter lasted for several hundred years, starting with Columbus' voyages (1492 onwards) between Spain and the Americas, which led to the Atlantic slave trade. It was this slave trade—the holocaust of *race-slavery*—that kick-started the story of 'race' in earnest. And then during colonialism the idea of the 'Other' became highly charged

© The Author(s) 2017
S. Fernando, *Institutional Racism in Psychiatry and Clinical Psychology*, Contemporary Black History, DOI 10.1007/978-3-319-62728-1_2

with notions of difference—*racial difference*—that stabilised racism as a powerful force in Western culture.

2.1 Race Thinking

Jacques Barzun (1937), a scholar of (mainly) history based at Columbia University, coined the term 'race thinking' in his book *Race: a Study in Superstition* and went on to state in the second edition of the book (Barzun 1965) that it 'rests on abstraction—singling out traits that are observed accurately or not, in one or more individuals, and making of these traits, a composite character which is then assumed to be uniform, or at least prevailing, throughout the group' (front flap of book cover). In other words, race thinking is the tendency to think of people in terms of physically and/or culturally recognisable groups (rather than individuals) without the sense that *all* individuals vary in physical appearance and, more importantly, in a wide range of psychological characteristics, cultural backgrounds and so on. As the notion of race became established, the ideology of *racism* (inherent within 'race') came to the fore. Essentially, racism is a way of thinking that places a superior *white people* in a position of power over *racially* inferior peoples of various other races—non-white races being delineated into a variety of 'Others', mainly on the basis of perceived skin colour, black, red, yellow, brown and so on. The notion of 'race' became established as a powerful sociopolitical force, seen as representing biological difference between human beings, during the era of race-slavery (the Atlantic slave trade), and the periods of Jim Crow in the USA and colonialism—all described in the next few paragraphs.

2.2 Exploration, Colonialism, Race-Slavery

The Middle Ages are sometimes called 'the Dark Ages' (with respect to most of Europe) because they were characterised by superstition, ignorance and economic stagnation. But in the Islamic Empire, which stretched across North Africa into what is called the Middle East (from a European perspective) or Western Asia (from an Asian or African one), and extended into Continental Europe through southern Spain (Arabic *Andalus*) to the borders of France, the Middle Ages were a time of cultural and economic development. In the fifteenth century, as Christian armies pushed the Islamic forces back across Spain, the Spanish Inquisition set up by King Ferdinand and Queen Isabella (Kamen 2014) picked on Moors and Jews as specific groups for persecution; and the defeat of the Moors by Christian forces at the battle of Granada in January 1492 led to the large-scale forced conversion and expulsion of Jews and Muslims, carried out instituted by the Spanish government with the help of the Inquisition (see Fig. 2.1). The year 1492 was also the year when Columbus sailed westward from Europe to explore the world beyond the Atlantic Ocean. Searle (1992) quotes evidence that Columbus had a map—possibly a 'Chinese map of the Americas' (Menzies 2008, p. 70)—that showed the sea route across the Atlantic to a land (eventually called America) where the people had much gold—rich pickings for anyone who could capture it; and that King Ferdinand and Queen Isabella of Spain, with the approval of the Pope, used the Inquisition to extract 'large amounts of money from Spanish Jewry' (p. 70) to fund Columbus' Atlantic voyages. Jan Carew (1992), a Guyanese writer and novelist and Professor Emeritus at Northwestern University, Illinois, writes: 'At the beginning of the Columbian era [when Spanish forces had subdued the Moors], thousands of books that the Moors had collected over centuries—priceless masterpieces that their geographers, mathematician, astronomers, scientists, poets, historians and philosophers had written, and tomes that their scholars had translated— were committed to bonfires by priests of the Holy Inquisition. And to cap this wanton destruction, an estimated three million Moors and 300,000 Jews were expelled from Spain (and this does not include the thousands forced to convert to Catholicism)' (1992, p. 3) (see Fig. 2.1).

Once they arrived in the Americas, the Spanish 'embarked upon a shameful course of ethnocide against indigenous peoples of the Americas that made its atrocities against conquered Moors, Jews and Guanches [aboriginal people of the Canaries who had been enslaved by the Spanish— see Searle 1992] pale by comparison' (Carew 1992, p. 5). Other European powers too arrived on the scene and conquest of indigenous peoples of the New World often proceeded with organised genocide when these peoples resisted forced labour, and the destruction of highly developed civilisations —described by Stannard (1992) as an 'American Holocaust' (1992). Thus, European ways of thinking about the 'Other', with their roots in anti-Semitism coupled with anti-Muslim attitudes (both deeply embedded in Europe at the time), set the stage for the much wider ideology of racism —one that objectified types of people sees as racially inferior (see 'Racialisation' in Chap. 5). Searle (1992) states that 'racism begat colonialism, which begat imperialism' (p. 70). The Portuguese moved enslaved Africans to Brazil from 1570 until 1630 when the Dutch took over control

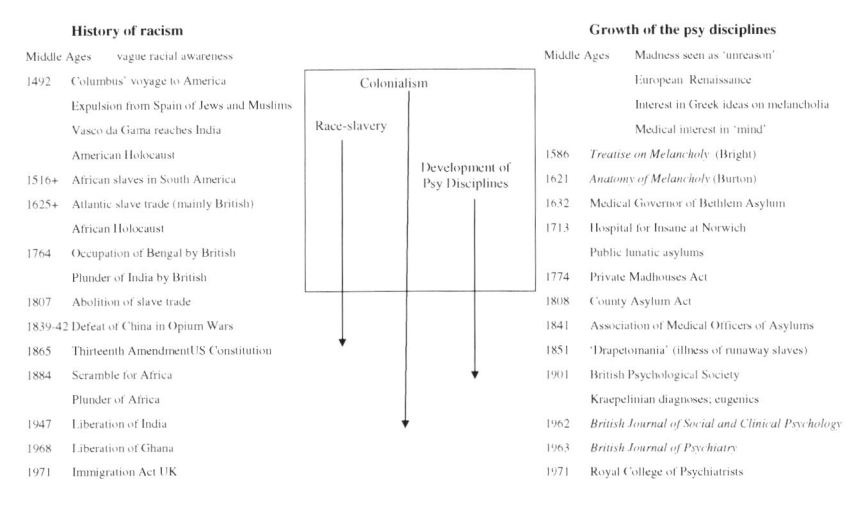

Fig. 2.1 Historic context of psychiatry and psychology

History of racism		Growth of the psy disciplines
Middle Ages — vague racial awareness		Middle Ages — Madness seen as 'unreason'
1492 — Columbus' voyage to America	Colonialism	European Renaissance
Expulsion from Spain of Jews and Muslims		Interest in Greek ideas on melancholia
Vasco da Gama reaches India	Race-slavery	Medical interest in 'mind'
American Holocaust		1586 — *Treatise on Melancholy* (Bright)
	Development of Psy Disciplines	
1516+ — African slaves in South America		1621 — *Anatomy of Melancholy* (Burton)
1625+ — Atlantic slave trade (mainly British)		1632 — Medical Governor of Bethlem Asylum
African Holocaust		1713 — Hospital for Insane at Norwich
1764 — Occupation of Bengal by British		Public lunatic asylums
Plunder of India by British		1774 — Private Madhouses Act
1807 — Abolition of slave trade		1808 — County Asylum Act
1839-42 — Defeat of China in Opium Wars		1841 — Association of Medical Officers of Asylums
1865 — Thirteenth Amendment US Constitution		1851 — 'Drapetomania' (illness of runaway slaves)
1884 — Scramble for Africa		1901 — British Psychological Society
Plunder of Africa		Kraepelinian diagnoses; eugenics
1947 — Liberation of India		1962 — *British Journal of Social and Clinical Psychology*
1968 — Liberation of Ghana		1963 — *British Journal of Psychiatry*
1971 — Immigration Act UK		1971 — Royal College of Psychiatrists

of sea access to South America. The large-scale transport of enslaved Africans —the Atlantic slave trade—began in 1625 in order to provide cheap labour for the plantations in the USA and the islands in the Caribbean, and was a massive project, run mainly by British companies (Walwin 1993) (see Fig. 2.1).

Millions of black Africans were forced onto ships and treated like cargo— as the personal possessions of their owners—transported to satisfy the demand for labour in colonies in America. During the 'middle passage', the hazardous voyage across the Atlantic, black people were seen and treated as savages by the slave traders. In the eyes of the local slave-owners in America, they were different to the indigenous people, (so-called) Indians. In *White over Black*, the classic book describing American attitudes toward enslaved Africans, Winthrop Jordon (1968) writes: 'Conquering the Indian [indigenous American] symbolized and personified the conquest of the American difficulties, the surmounting of the wilderness. To push back the Indian was to prove the worth of one's own mission, to make straight in the desert a highway for civilization. With the Negro it was utterly different. … And *difference,* surely was the indispensable key to the degradation of Negroes in English America' (p. 91, emphasis in original). Winthrop reckons that black people's status as relatively helpless strangers in America, together with their 'heathen condition' and blackness of complexion, set them apart from all other groups of people in Americas. There ensued a 'cycle of degradation' which, once established was accepted as a normal condition; and '[by] the end of the seventeenth century in all the colonies of the English empire there was chattel racial slavery of a kind which would have seemed familiar to men [and women] living in the nineteenth century' (Winthrop 1968, pp. 97–98). This was slavery, *racial slavery*, on an industrial scale that was different to slavery anywhere else before or since then—black people were slaves totally owned by their white masters. Most 'mixed-race' people were absorbed into either the black or white category (depending on appearance) while a few remained problematically in-between.

Slavery was virtually abolished 'by one means or another throughout the North' by 1830 (Woodward 1974, p. 17) and legally abolished in the whole of the USA in 1865 by the Thirteenth Amendment to the US Constitution: Abolition of Slavery (1865). But the freedom of black African Americans was circumscribed in many ways, the sum of which became known as the Jim Crow system (described in detail by Woodward 1974). According to Alexander (2012) Jim Crow was 'a term apparently derived from a minstrel show character' (p. 35). Jim Crow meant enforced segregation and legalised discrimination backed up (in the South) by illegal activities, such as acts of terrorism and lynchings carried out by the Ku Klux Klan; at the same time, the criminal justice system was strategically employed to force African Americans to live under a system which subjected them to extreme repression and control (especially vicious in the Southern states—'a tactic that would continue to prove successful for generations to come' (p. 32). The extreme version of Jim Crow was gradually eroded by migration to the North, and the increasing political organisation among black people that led to the civil rights movement following the end of WWII; such organisation resonated with the liberation movements that were gathering pace in other parts of the world as white-supremacist colonialism was overthrown (see 'Transformations after WWII' in Chap. 4). It was during the era of racial slavery in the USA, and then during that of Jim Crow, supplemented by the colonial projects in America, Africa and Asia, that the stage was set and the principles established for relationships between white people and 'other' races that were characterised by racism and underpinned by the ideology of white supremacy, and which continue to the present day—not just in the USA, although that is where these are seen most vividly, but all over the world.

In the early part of the fifteenth century, Portuguese explorer Vasco de Gama, helped by Arab sailors, found the sea route to India round the southern tip of Africa and across the Indian Ocean, thus opening up to Europeans the sea routes to the East (the southern silk road) previously monopolised by Arab and Chinese traders (Hall 1996). These routes were first exploited by the Portuguese and Spanish but soon afterwards Great Britain, the Netherlands and other seafaring nations joined in, not just for purposes of piracy and to search for wealth and lands to which their people could emigrate, but to explore and to discover exotic places, people, fauna and flora—although admittedly the funding for such journeys was often provided on the (often explicit) understanding that lands allegedly 'discovered' would be claimed for the European mother country. Exploration soon became exploitation, leading to colonialism and race-slavery (or the semi-slavery of indentured labourers) of people 'different' to the colonists; and, where circumstances permitted, Europeans were brought over for settlement—sometimes on a very large scale—on lands that had been appropriated from people of 'lower' races, who were pushed out or even subject to genocide. Natural resources were seized one way or the other (for example by unequal treaties); local skills were exploited (and sometimes copied for use in the mother country in Europe); and land and property were obtained by unequal trade agreements often imposed by force of arms. Most importantly, local people were tricked or forced into servitude and, almost from the start of European contact with Africa, vulnerable inhabitants were enslaved and transported for forced labour—in small numbers to Europe and European colonies in Asia, and in large numbers, in the notorious Atlantic slave trade, to the European colonies in America. Aggressive colonisation, the transport of enslaved Africans to work in the New World and the plunder of the other continents by European powers would characterise the next 400 years—something beyond the remit of this chapter to describe in detail. After the rise of racism (which, as we saw earlier, was necessary to justify the enslavement of Africans on an industrial scale) during the Atlantic slave trade, 'European thinkers were concerned to keep black Africans as far as possible from European civilization' (Bernal 1987, p. 30).

As India was gradually controlled, plundered and colonised (a process comprehensively described by Shashi Tharoor 2016), European nations attempted to trade with China. They had little to offer in exchange for the tea, valuable fabrics and other consumer goods that China possessed, until Britain had the idea of exporting opium grown in (British) India to China. As European nations mounted attacks on China during the nineteenth century, political and social structures in the country were destabilised

(Bernal 1987). China was forced by the opium wars to accept so-called 'free trade'—a system that allowed the British and French to engage freely in importing and selling opium on the Chinese mainland. Consequently, the balance of trade between China and Europe shifted in favour of the latter in the nineteenth century. China became a virtual client state (of Western powers), semi-colonised, with warlords and factions dependent on European powers. Egypt, like China, had a long and illustrious history of civilisation and culture, going back much further than anything the West had to offer; and ancient Egypt, which in the eighteenth century had been seen as a very close parallel of China, suffered a similar fate to that country in the course of European expansion into North Africa (see Fig. 2.1).

2.3 The European Enlightenment

The period of cultural changes in eighteenth-century Europe, usually referred to as the 'Enlightenment', is often thought of as significant in the development of European thinking that led to the development of what today are sometimes called the 'European values' of liberty, democracy and equality. Dorinda Outram (2005) Professor of History at the University of Rochester argues that the Enlightenment was not a single process or movement but 'rather a capsule containing a set of debates … characteristic of the way in which ideas and opinions interacted with society and politics' (p. 8). These ideas and opinions included only those emanating from Europe and the key thinkers (or major figures) who set the tunes were all white Europeans. Outram writes: 'The Enlightenment relies on "rationality", reasoning which is free from superstition, mythology, fear, and revelation, which is often based on mathematical "truth", which calibrates ends to means, which is therefore technological, and expects solutions to problems which are objectively correct' (p. 6). But that was not the whole story. David Goldberg (1993) points to the Enlightenment as a highly racialised project: 'A few examples will suffice. Kant citing with approval David Hume's likening of learning by "negroes" to that of parrots, insisted upon the natural stupidity of blacks. John Stuart Mill, like his father, presupposed non-white nations to be uncivilized and so historically incapable of self-government. Benjamin Disraeli captured the sensibility of the mid-nineteenth century by declaring the only truth to be that 'all is race". The basic human condition—and so economic, political, scientific, and cultural positions—was taken naturally to be race determined' (1993, p. 6).

The philosopher Emmanuel Chukwudi Eze (2001) has carefully explored the shift in European thinking that took place in the age of the Enlightenment: 'For the ardent Cartesian [an adherent of ideas enunciated by the French philosopher René Descartes, dubbed the father of modern Western philosophy] … human differences of skin color or sex are merely accidents, inessential, and illegitimate criteria for determining the essential and true worth of the person'; but the empiricism in the writing of Hume and other philosophers of the Enlightenment dictated that human essence was in the body—'Hume, unlike Descartes, denied the existence of a metaphysical essence of human nature'. Eze writes that Harry Bracken, Professor of Philosophy at McGill University and a specialist on Descartes, 'found an intriguing historical correlation between the rejection of Cartesianism in England and the simultaneous growth in that country of empiricism, colonialism, and racism' (2001, p. 54).

The Enlightenment was the time when liberalism—the 'tradition of thought whose central concern is the liberty of the individual' (Losurdo 2014, p. 1)—emerged in Europe. Yet, at the time the Atlantic slave trade, almost a monopoly of British companies, was in full swing; the plantations in the USA and the Caribbean on which chattel slavery flourished were predominantly in British colonies; and ironically, liberal ideas were being pursued by English gentry and plantation owners (Losurdo 2014). The American writer Toni Morrison (1993) points out: 'The concept of freedom [during the Enlightenment] did not emerge in a vacuum. Nothing highlighted freedom—if it did not in fact create it—like slavery. … What rose up out of collective needs to allay internal fears and to rationalize external exploitation was an American Africanism—a fabricated brew of darkness, otherness, alarm, and desire that is uniquely American' (1992, p. 38). This history resonates today in the current experiences of African-Americans (see the section 'Obama years' in Chap. 8).

The (European) Enlightenment may be overrated in its significance, considering that it only affected a minority of people in the world and a few cultural groups, but the changes in Europe certainly set the stage for many cultural and political shifts throughout the world in subsequent years, partly because of European domination of Asia and Africa during the following 300 years or so; and the 'values' of the Enlightenment are quoted in English literature and European politics as those that people everywhere should aim at achieving. The Enlightenment thinkers' legacy in inspiring the French revolution may have enabled the first black republic outside Africa to be formed by a rebellion by the enslaved people of Haiti, who overthrew their French masters; but the liberal ideas of the Enlightenment at the time they were first propounded were race-specific, and only applicable to white people. The writings on race by eminent European thinkers of the Enlightenment—according to Eze (1997) 'Hume, Kant and Hegel played a strong role in articulating Europe's sense not only of its cultural but also *racial* superiority (p. 5, emphasis in original). And the American Africanism referred to by Toni Morrison—see above—with its counterpart, an European Africanism seen in colonial literature and art (for examples see Smith 2015), created the myths and stereotypes of what people who were seen as 'African'—black people—represent today in (white) European culture. It must be noted that in searching for knowledge and understanding of the world and the human condition, 'the problems we pose, the theories we use, the methods we employ, and the analyses we perform are social products themselves and to an extent reflect societal contradictions and power dynamics' (Bonilla-Silva 2014, p. 13).

2.4 Scientific Racism

In the eighteenth century, Swedish botanist Carl Linné, generally known as Linnaeus (1758–9), who devised a formal system of naming species of living things, extended his classification of plants and animals to divide human beings according to a hierarchy development based on skin colour —whites at the top. Physical anthropology developed methods for classifying skulls to indicate levels of intelligence—again with Europeans at the top (Jordan 1968). The anthropological and medical contention that the brains of black people were inferior to those of whites was supported by dubious research in the nineteenth century and early twentieth. For example, Robert Bean (1906), Professor of Anatomy at Johns Hopkins University concluded: 'From the deduced differences between the functions of the anterior and posterior association centres and from known characteristics of the two races the conclusion is that the Negro is more objective and the Caucasian more subjective. The Negro has lower mental faculties (smell, sight, handcraftmanship, body-sense, melody) well developed, the Caucasian the higher (self-control, will-power, ethical and aesthetic senses and reason)' (Bean 1906, p. 412).

To demonstrate the (pseudo-) scientific racism of the time, Fryer (1984) quotes the following tenets of anthropology, summarised by Hunt (1863) in his presidential address to the Anthropological Society of London (which Hunt founded), in words that document the thinking of the times:

1. That there is good reason for classifying the Negro as a distinct species from the European, as there is for making the ass a distinct species from the Zebra: and if, in classification, we take intelligence into consideration, there is a far greater difference between the Negro and European than between the gorilla and chimpanzee. 2. That the analogies are far more numerous between the Negro and the ape than between the European and the ape. 3. That the Negro is inferior intellectually to the European. 4. That the Negro becomes more humanised when in his natural subordination to the European than under any other circumstances. 5. That the Negro race can only be humanised and civilised by Europeans. 6. That European civilisation is not suited to the Negro's requirements or character.

(Hunt 1863, cited by Fryer 1984, p. 177)

Although scientific racism in the mid-nineteenth century seemed to have sealed a narrow biological view of race into European culture, major changes were to take place in the understanding of 'race' and the ways in which racism impacted on people and social systems (see Chap. 4).

2.5 Origins of Western Psychology and Psychiatry

The disciplines of psychology and psychiatry that arose in European culture are referred to in several places in this book as 'psy' disciplines in order to emphasise how closely they are bound up together, both in terms of their histories and their function in Western society and, increasingly, outside the West. Both disciplines are involved, both in clinical work with clients and for patients attending 'mental health services' that are regarded as part of *medical* services. Critical thought around matters to do with mental health tend to focus on the 'psy' disciplines because of the power they wield by (a) having a major input into labelling of people in terms of their social functioning, capacity to be responsible for their behaviour, personality and mental states—in particular the propensity to be dangerous to, or a burden on, others; and (b) underpinning the style of how a variety of sociopolitical systems (especially those providing the public with mental health services) are structured. Matters 'mental', many of which used to be thought of as within the purview of 'religion', are now thought of as issues of 'health' (often public health)—a conflation that is particular to Western culture but one that is increasingly being spread worldwide as local non-Western systems are being globalised (Fernando 2014a, b). Apart from underpinning ways of working in mental health services, the 'psy' disciplines influence the functioning of health services in general. And they greatly influence the training of professionals in the mental health field, whether as psychologists, psychiatrists, mental health nurses, social workers, occupational therapists or others involved in service provision. Both disciplines are founded on the academic study of psychology and psychiatry (sometimes called medical psychology) and both arose in tandem within the general field of Western knowledge.

Gardner Murphy (1938) traces the beginnings of modern academic psychology (which led to clinical psychology) to the revival of learning in Europe during the fifteenth and sixteenth centuries (the so-called Renaissance) leading to that period of European history in the seventeenth and eighteenth centuries referred to as the Enlightenment, the age of reason (Barzun 2000; Smith 2008). In the sixteenth and seventeenth centuries there was a growing interest in melancholia, derived from Greek literature (possibly through its elaboration in the Arabic writings of the tenth to the thirteenth centuries); this interest was represented by books such as *Treatise in Melancholy* by Timothy Bright (1586) and *Anatomy of Melancholy* by Robert Burton (1621) (see Fig. 2.1). Prior to the Enlightenment the Bible was the fount of all knowledge and the Western world was in effect theocentric, deriving its (supposed) authority from Christian teachings of the time. European thinking about the human condition moved during the Enlightenment from the medieval, theocentric 'Dark Ages' (now seen as largely dominated by 'superstition') to the modern, body-centred world of the self-made person. The process that enabled this to happen has been described by Porter (2004) as the 'psychologizing of the Self' (pp. 347–373), and as amounting to a reinterpretation of identity in psychological terms—indeed a dismantling of what had been the centre of human condition, namely the 'Soul', concurrent with the exclusion of spirituality (represented in the West in religion). As much as a cultural shift, this change was made possible by the power struggle between science and the Roman Catholic Church (see below, 'The scientific paradigm') The psyche, which had originally meant the soul (with spirituality), something relatively active, came to be considered more as an inert, static 'mind'. And study of the psyche ultimately developed into psychiatry (concerned with the abnormal mind) and psychology (focused on the normal): the 'psy' disciplines.

The 'psy' disciplines became standardised and elaborated (in Western culture) through the study of people designated as being 'mad' in a context of Enlightenment thinking (Foucault 2006) and the next few paragraphs explore the pressures and context of the time within which this happened. However, some points (discussed at some length elsewhere, for example in Fernando 1991, 2014a) should be noted: first, madness had been recognised all over the world in many cultures (Porter 2002) and every cultural tradition includes a concept of illness (McQueen 1978). But concepts of 'mind' and mental functioning (which psychology was primarily concerned with) and notions about 'illness' of the mind and human behaviour (which was the focus of psychiatric knowledge) developed differently in non-Western cultural traditions from that in the West. Second, with regard to Western knowledge sources, some ideas in the 'psy' disciplines hark back to Greek thinking (for example, to Plato, Socrates and Hippocrates—see Fernando 1991, pp. 53–54); and Greek medicine helped promote a medicalised concept of madness that informed the work done in the *mãristãns*, or medieval hospitals for the insane in the Arabic Empire (Dols 1992, pp. 28–29; Foucault 2006, p. 117; Fernando 2010, pp. 48–51 and discussions in Fernando 2014a, pp. 28–29) although, with the decline of that empire, much of its knowledge, written up in Arabic, became corrupted and was lost to the knowledge base that developed within European cultures (see Fernando 2014a, pp. 28–29).

Two matters should be noted here, although they are explored further in other parts of this chapter and later chapters of the book. The first is about the 'objectivity' that a scientific approach should have ensured if Enlightenment values were to underpin the development of the 'psy' disciplines—the issue being that this is not what actually happened (see Fernando 2010, pp. 53–56; and also see the section, 'The scientific paradigm', later in this chapter). The 'psy' disciplines only *mimicked* objectivity; for example many of the methods they used for understanding individuals were clearly subjective, although packaged in language that suggested otherwise. As psychologist Thomas Graham (1967), reviewing critically the way (Western) medicine (of the body not the mind) differs today from clinical psychology with respect to objectivity, says: 'the tower of the *physician* stands unshaken whilst the temple of the *psychologist* [and psychiatrist] rocks to its foundations' (p. 41). The second matter of note is about racism. As the Western disciplines of clinical psychology and psychiatry became established in Europe and in Europeanised America, they reflected the values of the (European) Enlightenment in general, and racism and the ideology of white supremacy attached to racism were part of that Enlightenment's values (see the above section, 'The European Enlightenment').

2.6 The Scientific Paradigm

René Descartes, seventeenth-century French philosopher and mathematician, who is sometimes seen as the father of modern European philosophy, saw human activity as being either of a mechanical nature or resulting from rational thought. For Descartes, the soul (which is now interpreted as the 'mind') is purely spiritual and formed of a different substance to the 'body' (Koyré 1970, xliii); and the Cartesian doctrine that has come to us is the dogma of the Ghost in the machine ... that there occur physical processes and mental processes; that there are mechanical causes of corporal movements and mental causes of corporal movements (Ryle 1949, p. 23). Later in the seventeenth century, Newtonian physics came on the scene. The natural world became a mechanical system to be manipulated and exploited; living organisms were seen as machines constructed from separate parts, each part being broken into further divisions. And scientific study was necessarily reductionist. There emerged the view of mind as an objective 'thing' to be studied by objectified methods, the preferred approach being reductionist (see below under 'Biologisation of mind'). As the separation of natural philosophy from theology progressed in eighteenth-century Europe, science (linked to natural philosophy) to a large extent replaced 'religion' (linked to the teachings of the Church) as the main source of knowledge about the human condition—so much so that 'at the beginning of the century the most widely purchased books were theological; [but] by the end of the century they were fiction or popular science' (Outram 2005, p. 107). The power of the Church as keeper of 'religion' (in being the interpreter of the Bible) was replaced by the power of (scientific) observations and theories based on rationality. The resulting conflict between the Catholic Church and the scientific establishment is epitomised by the *Galileo affair* (Finocchiaro 1992)—the trial and condemnation of a scientist for his scientific views, which were seen by the Roman Inquisition as heresy. The stage was set for a 'scientific approach' that was part of the flowering of new ideas (the Enlightenment) in the eighteenth century.

The concept of a paradigm first drawn attention to by Thomas Kuhn, writing mainly about science, is a useful way for us to consider the nature of knowledge production—epistemology. It means a system of beliefs and assumptions that determines fact-gathering within a system of knowledge

—'the rules of the game', which are often implicit, rather than being clearly stated, more like shared beliefs than explicit theories or practical guidelines (Kuhn 1962, pp. 40–45). A paradigm determines 'what constitutes useful and respectable data', how the creators of knowledge 'go about their business' and so on (Ingleby 1980, p. 25). The basic features of the scientific paradigm that emerged from the ideas of the Enlightenment can be divided into beliefs and approaches (see Table 2.1). Beliefs centre on (a) positivism, the belief that reality is confined to what can be observed, and knowledge limited to events and to empirically verifiable connections to events; this means ignoring everything prohibited by the existing 'reality'—'that is everything that does not exist, but would under other conditions, be historically possible' (Martin-Baró 1994, p. 21); (b) causality, meaning that everything that exists has a cause (for its existence), which yields a mechanical cause-and-effect model and implies that nothing is truly random and nothing beyond understanding (supernatural); (c) objectivism, where feelings and experiences become things 'out there' to be studied as objects, and moral and ethical judgements are not valid; and (d) rationality, where the final arbiter of truth is reason and all assertions verifiable by logical reasoning. The methods of study in scientific thinking (at the time when the 'psy' disciplines developed) were (a) the mechanistic

Table 2.1 Scientific paradigm

Beliefs
Positivism
Reality is rooted only in what can be observed
Causality
Nothing occurs randomly
Natural causes for all events and effects
Objectivism
Feelings, thoughts etc. regarded as objects
Rationality
Reason superior to emotion
All assertions verifiable by logical proof
Approaches
Mechanistic
Newtonian physics as opposed to modern physics
Reductionist
Sum of the parts equals to whole
Logical reasoning
Intellectual exercise

approach of Newtonian physics; (b) the reduction of complex systems into their parts; and (c) intellectual, logical reasoning as opposed to any other type of understanding, such as intuition.

As in the case of the natural sciences (physics, chemistry and so on) the 'psy' disciplines that took root in European culture in the nineteenth century tried to do so within the scientific paradigm (Table 2.1). The natural sciences generally succeeded but Western psychology and psychiatry tended to fall far short (see the discussion of their troubled path in the section 'Sociopolitical context', below). For example, most of the concepts used in the 'psy' disciplines tend to depend on subjectivity and so lack scientific validity. Also, as a result of the political and theoretical conflict with the Catholic Church representing 'religion', anything to do with spirituality (seen as a part of religion) was excluded as unscientific. The exclusion of spirituality resulted in the 'psy' disciplines being secular in approach—something that has become a major drawback in the case of their *clinical* applications in psychiatry and clinical psychology, which are to do with human beings in all their varied dimensions, mind, body and spirit.

In following the scientific reductionist approach, general medical sciences of the nineteenth century examined smaller and smaller fragments of the human body, looking for pathologies to be measured as objectively as possible, while tying these up with how ill-health was manifested as complaints (constructed as symptoms), and relating these to (objectively measured) pathologies. The imperative of the 'psy' disciplines to follow— or seeming to follow—the scientific approach was strong; and so they followed the pattern set down by the general medical sciences—objectifying complaints and problems seen as not having a medical basis, attributing them to particular locations in the mind or interactions between different parts of the mind. Thus, a variety of problems of living, which in reality were complex emotional and social problems and problems of relationships and exigencies of life, were reduced to pathologies (in the psyche or mind) and thence to illnesses (interpreted as diagnoses based on the tradition of medicalisation of human problems, see Fernando 2014a, pp. 27–28) or formulations from a variety of theoretical perspectives (see Johnstone and Dallos 2014). The budding clinical psychologists and psychiatrists had to prove their worth as 'scientists' (in the case of psychiatrists there was additional pressure to be recognised as 'real doctors' like their counterparts in medical specialities) who looked to objective tests to validate pathologies.

The result of the struggle of those in the 'psy disciplines' to be accepted as scientific practitioners was that their academic approaches (often using complex statistics) were presented as objectively measured matters rather than what they were (and are): products of social construction, introspection, inspiration, imagination and guesswork. The study of the variety of concepts they dealt with, for example parts of the mind, such as the ego, the unconscious and intelligence; pathological processes in mental illness; and diagnoses like neurosis and melancholia, were thought to be probably valuable in some circumstances, but were not really considered as 'science'. However, by adhering to observation (however biased it may be), experimentation where possible using the soft data of 'observation' and the quantitative processing of data, the 'psy' disciplines have generally been able to claim a scientific status equivalent (in public perception) to that of many medical sciences, if not the hard sciences such as physics and chemistry.

2.7 Biologisation of Mind

In the latter part of the nineteenth century, a French doctor, Bénédict-August Morel, preoccupied by 'the seemingly remorseless rise in the numbers of the insane and the apparent inability of mental medicine to cure its patients' proposed the concept of 'degeneration'—deviation from normality recognisable by physical stigmata that resulted in the moral and intellectual collapse of an individual (Pick 1989, p. 54)—as an explanation for madness. This led to the notion of madness being inherited and the assumption of the inherited nature of much of what constitutes human psychology. At the same time, the notion of race played an important part in determining how both madness and human psychology (the understanding of the mind) were seen. As the 'psy' disciplines became influenced by social Darwinism, psychology espoused eugenics, 'the science which deals with … inborn qualities of race' (Galton 1904, p. 1) and 'race psychology' became popular, especially in the USA—see 'Nineteenth-century psychology and psychiatry' and 'Race psychology' in Chap. 3. The mind was increasingly seen as something produced (biologically) by the brain, and illness of the mind as essentially biomedical.

A parallel movement to biologisation that had a deep impact on the 'psy' disciplines was the rise, during the first half of the twentieth century, of Freudian psychoanalysis. Freud himself was a neurologist and medical doctor who expected that the psychological theories he propounded would eventually be substantiated *materially* in neurological systems located in brain anatomy; but the importance of psychoanalytic theories was that they argued for a psychic (or functional) basis for brain activity and hence an alternative to the search for a materialistic anatomic-biological basis for the 'psy' disciplines. Erich Fromm saw psychoanalysis as an attempt to bring an element of spirituality into the 'psy' disciplines (see Fromm et al. 1960), but the impression made by psychoanalysis collaborated with the forces of the biologisation of the mind to exclude spirituality from both (Western) psychiatry and psychology. At the turn of the nineteenth into the twentieth century, the so-called illness 'schizophrenia' was constructed as the epitome of genetic illness of the mind. By the middle of the twentieth century, all mental disorders were viewed as inborn conditions, which *ipso facto* (at that time) were not amenable to treatment. Since the end of WWII in 1945, and even more so since the medication revolution of the 1970s, the 'psy' disciplines have gained in prestige and power in much of the world (Fernando 2014a). Their basic tenets, set in the eighteenth and nineteenth centuries, have not shifted much at the level of practical work, that is clinical work with people deemed to suffer from mental health problems, although there are movements in the academy questioning these tenets—such as the 'critical psychiatry network' (http://www.criticalpsychiatry.co.uk/) and 'madness studies' (http://madness-studies.com/). Today, in the second decade of the twenty-first century, the belief in reified concepts like 'schizophrenia' is an important part of the belief system of the 'psy' disciplines—and a major part of the paradigm that informs them.

2.8 Sociopolitical Context

Prior to the (European) Enlightenment (see above), madness was seen as the state of mind of people who had not merely lost their 'reason' or did not possess it to start with, but people who possessed 'unreason' (Foucault 2006), a somewhat positive condition or even ability. As some classes of people in Europe became rich through the proceeds of slavery and the trade in sugar, and the plunder of their colonies, the growth of cities and of affluent areas of residence, and thereby disparities of wealth and health (in Europe) became evident, and some groups/classes of people designated as 'mad' were marginalised and seen as an encumbrance, even a danger. People considered socially undesirable were segregated and placed in institutions, well away from 'respectable' people. The building of asylums for the 'mad' was initially limited to Europe but then spread across to

European settler colonies (in for example, America and Australia) and was later imposed on some Asian and African colonies by imperial powers (see Fernando 2014a, and 'Alleged mentality of black people' in Chap. 4). Asylum inmates (as they were called) were placed there partly for their own protection and care, and partly to protect the general public from the insane. As the mad became a burden on society at large, the concept of unreason as a positive state, and not just as the lack of reason, gradually lost its significance; and, as the mad fell under the domain and control of medical doctors, unreason was increasingly seen as an 'illness'. Being locked up, they (the mad) became the object of study and observation—leading to the development of a system of knowledge that gave us the discipline of psychiatry and its counterpart, clinical psychology (Fig. 2.1).

By the end of the eighteenth century, lunatic asylums had been largely renamed 'mental hospitals', and the need to 'cure' went hand in hand with the purpose of protecting society: 'The mad are now [eighteenth century onwards] locked up in *order to be cured*' (Khalfa 2006, p. xviii, emphasis in original). As the asylums became central to the development of the mental health professions, and (medical) doctors were now in charge of asylums, psychiatrists (who were medically qualified) achieved a higher standing (in comparison to clinical psychologists) in the field of mental health. What happened in practice was that psychologists tended to be restricted to making a psychological (including IQ) assessment, while psychiatrists took on the roles of both assessment and treatment, and of being responsible for predicting likely outcomes (prognosis)—and this gave psychiatrists, and hence the discipline of psychiatry, greater prestige and legal power. However, from the second decade of the twentieth century onwards, clinical psychologists would be increasingly involved in all aspects of mental health care and today both disciplines are seen as more or less equivalent. Both underpin the theory and practice of what happens in mental health services.

In most of Europe, doctors in charge of asylums were once called 'alienists'—people who decided who was alien to (respectable) society and who was 'normal', madness itself being sometimes referred to as 'mental alienation' (Shorter 1997, p. 17). By the end of the nineteenth century, the quest to understand 'unreason' (the original characteristic of madness) had all but disappeared, although parts of Freudian psychoanalysis in the early part of the next century 'raised the possibility of a dialogue with unreason', albeit within a medical framework (Foucault 2006, p. 339). Alienists used various diagnoses depending on the sociopolitical context.

Racism infiltrated diagnoses from the very start— the most (in) famous, in early nineteenth century, being drapetomania (the illness characterised by running away), one of several diagnoses given to Africans who protested the condition of slavery at the time (see 'Mental pathology and the construction of race-linked illnesses' in Chap. 3). A standard for a diagnostic classification on mental illness was pursued by alienists (see Fernando 2014a); and the model proposed by Emil Kraepelin (1896, 1913) at the turn of the nineteenth to the twentieth century—sometimes called the Kraepelinian or Kraepelian approach (Donald 2001)—became dominant and still remains the basis of modern (biomedical) psychiatry (see Fig. 2.1).

In the UK, alienists (often called 'mad doctors') came together in the Association of Medical Officers of Asylums and Hospitals for the Insane, founded in 1841; its *Asylum Journal of Mental Science* was first published in 1853. The Association became the Royal Medico-Psychological Association, and finally the Royal College of Psychiatrists (RCP) in 1971, publishing the *British Journal of Psychiatry*. The British Psychological Society (BPS) evolved from The Psychological Society, founded in 1901 at University College, London. The BPS publishes many journals, one of which is devoted to clinical matters, the *British Journal of Clinical Psychology*. Figure 2.1 shows how the context in which psychiatry and clinical psychology developed was one in which race thinking was the norm and racist ideologies ruled supreme, having been fashioned by race-slavery and colonialism. Considering the extent to which context influences social and cultural structures, it was always likely that the 'psy' disciplines would be institutionally racist—unless the disciplines themselves recognised this and took action to counteract racism.

Once madness was seen as an 'illness', it was seen as something separate from the 'mad' person, to be dealt with *positively*, on the basis of positive facts about illness—for example, each illness had a 'natural history' and verifiable data and so on—in keeping with what was then the scientific approach. As the 'psy' disciplines became the body of expertise about mental illness, they established power over the mentally ill and also made it possible for the 'appearance of a psychology [of mental illness, of madness and thence of normality] ... a cultural fact peculiar to the Western world since the nineteenth century' (Foucault 2006, p. 529).

The two 'psy' disciplines thus worked as partners interested in the field of mental health, psychology being concerned with accumulating knowledge about the (supposed) functioning of the (normal) 'mind'; while 'psychopathology', the supposed pathologies of the mind—the abnormal mind—was the province of psychiatry; and both disciplines attempted to operate within a scientific paradigm developed in the eighteenth and nineteenth centuries (see Table 2.1). The adherence to a scientific approach in the 'psy' disciplines seemed justified at the time because such an approach seemed to be paying off in the case of biological therapies in medicine and in the practical applications of that approach in chemistry and physics. However, it should be noted that modern science (the 'new physics') has a different paradigm to the nineteenth-century scientific paradigm (which informs the 'psy' disciplines); for example in the case of Heisenberg's uncertainty principle, and chaos theory, suggesting the importance of unpredictability (see Davies and Gribbin 1991).

Psychiatry and clinical psychology have pursued a troubled path in the field of mental health because of the need for mental health professionals to understand problems of living and the *lived experience* of human beings, who are *subjective* beings: this was difficult, because these could not be subsumed easily within the objectivity predicated by a scientific approach (Table 2.1). Clinical psychology itself struggles to maintain a balance between the various influences it is subjected to. As Foucault (2006) states it remains 'by its very nature, at a crossroads ... between the subject and the object, between within and without, between lived experience and knowledge' (pp. 529–530). And the same could be said for (clinical) psychiatry, although its political adherence to the structures of medicine (the illness model) means that it is pulled towards objectivity even more strongly than clinical psychology. The struggle of both 'psy' disciplines, as *clinical* disciplines (rather than purely academic ones) to keep to a middle path between subjectivity and objectivity, has meant that, both the institutions and the individuals working in them (as psychiatrists and clinical psychologists) have, from the beginning, been split into two broad camps —the sociocultural and the biological. This split is reflected even today in the UK within the official (professional) bodies, the Royal College of Psychiatrists and the British Psychological Society, especially in its the latter's Division of Clinical Psychology; and it is likely to be exacerbated in the future (see 'Future of the "psy" disciplines' in Chap. 9).

2.9 Limitations of Knowledge

Since the disciplines of psychiatry and clinical psychology developed in a context of Western, and to a large extent, *West-European*, cultures, drawing little, if anything, from 'other' cultural traditions, the question arises as to their suitability *culturally* for informing mental health services meant for people whose backgrounds may not be culturally *Western*. This is a complex topic that is beyond the scope of this book, although some indications may be apparent when the place of racism and other forms of discrimination in such diagnoses are considered in subsequent chapters. However the next few paragraphs will make some points that suggest the cultural limitations from which the 'psy' disciplines suffer because of their historical failure to draw on knowledge systems in non-Western cultures. In most non-Western cultural traditions, including those of pre-Columbian America, the conceptualisation of mind and body and ideas about illness and health developed very differently to those in the West.

A major problem in discussing non-Western cultural forms vis-à-vis 'mental' matters is that a reliable body of information on the background and traditions of Africa and pre-Columbian America is not available for several reasons: in the case of Africa, the subject is vast and relatively unresearched (Karenga 1982); and the keepers of historical knowledge in the past were mainly griots, 'professional oral historians' (p. 53). Even more importantly, European conquests led to the loss of information about African societies and what was collected as history was often distorted to fit into racist models of African 'primitiveness'. In the case of South America, the wanton destruction by the Spanish conquerors that followed Columbus resulted in genocide, plunder and cultural pillage, all but destroying the civilisations in the region that preceded European conquest. In North America, aggressive colonisation by Europeans left the indigenous people restricted to life in reservations, and thereby destroyed their traditional cultures (Haig-Brown 1988). Thus deductions in non-Western cultural forms of what is equivalent to (what in Western cultures may be referred to as) 'the mind' are not easy to describe in a short section—there is a vast literature on this topic and some explanations have been made in chapters of books I have written (Fernando 2002, 2010 and Fernando and Moodley, in press). All that can be stated is that, from what we can decipher, the understandings of the individual mind in Asian and African psychologies are very different in fundamental ways to those in Western psychology and psychiatry. Not only are the roots of non-Western ideas very different—for example, their concepts of mind did not come out through a study of madness but from a mixture of spirituality and personal introspection—but so are the ways in which they relate to concepts equivalent to Western ideas of health and illness. Even today, non-Western psychologies remain to a great extent embedded in religion and philosophy, unlike Western psychology, which aims to be 'scientific' (or whatever goes for 'science' today) and allied to (Western) medicine.

Another way of looking at the cultural limitations of the Western 'psy' disciplines is to examine what else was going on historically in the field of what is now known as 'mental health' at the time of their development; that is, the knowledge that those 'psy' disciplines did not draw on. A clear approach to madness was evident in the practices within the *māristāns* of the medieval period (which flourished from the tenth to thirteenth centuries) described by Dols (1992). According to this approach, the underlying theory of illness was seen in the humoral terms of Greek medicine, which was elaborated in a Islamic-Arabic context and possibly influenced by ideas from Hebrew culture—the Hebrew scholar and Jewish rabbi Maimonides (whose statue still graces a square in Cordoba) was one of the main writers on mental illness at the time. Foucault (2006) states that, unlike in European institutes of the times, 'a sort of spiritual therapy was carried out [in the Islamic hospitals], involving music, dance, and theatrical spectacles and readings of marvellous stories' (2006, p. 117). According to Graham (1967) Islamic 'psychiatry' (if we can call it that) encapsulated 'a blissful union of science and religion' (p. 47). Tibet, as a landlocked and geographically isolated place, developed a system of medicine that was (and is) unique, different to both Indian and Chinese medical systems, although drawing from both. The system of healing for madness within Tibetan medicine has been dubbed 'Tibetan psychiatry' by Clifford (1984). She calls it a 'psychiatry' because it combined the teachings of orthodox Buddhism, or rather the Tibetan Buddhist elaboration of these, with the application of ideas of herbal therapy and diet derived from Ayurveda (one of the main Indian systems of medicine), to form a method of person-centred treatment for mental and emotional problems, including madness: 'A complex interweaving of religion, mysticism, psychology, and rational medicine' (p. 7). Clifford uses psychoanalytic images to describe how psychosis was seen from a Tibetan medical perspective: problems could build up into 'a tremendous panic … [associated with] … repression that is elaborated in terms of ego and unconscious tendencies … eventually leading to psychosis' (p. 138). In the Tibetan system, '[T]he psychological basis of insanity is the same basis for enlightenment. It all depends on whether or not it is accepted and comprehended and ultimately worked with as the key to liberation' (pp. 138–139).

2.10 Modern Psychiatry and Clinical Psychology

The embedding of racism in the 'psy' disciplines forms much of the discussions in several chapters of the book. As the earlier 'Sociopolitical context' section of this chapter makes clear, the 'psy' disciplines were influenced by sociopolitical forces, throughout their development from the study of madness and the control of people seen as mentally ill—one could in fact argue that these forces were all-important. Although they have undergone changes over the years, these disciplines' fundamental approach stays the same, continuing to reflect the cultures and sociopolitical contexts in Euro-America at the time they developed. The interface between the West and the Rest from 1492 onwards—the Dark Ages referred to by Kapuściński (2008) (see first paragraph of this chapter)—are depicted in Fig. 2.1 as the 'history of racism'. Clearly, it was inevitable that racist notions would permeate the 'psy' disciplines as they developed (as shown on the right-hand side of Fig. 2.1) unless definite action was taken to prevent that happening.

The practice and organisation of mental health services in the West (Europe and North America) have undergone quite significant changes since the end of WWII. Almost mimicking the great confinement that dominated the Europe and America of the eighteenth and nineteenth centuries (when the asylum movement arose), the decades between 1960 and 1990 saw the emergence of a drug-based psychiatry, drugs being the main line of treatment because of the assumption that mental illness was caused by chemical imbalance in the brain—I have referred to this in an earlier book (Fernando 2014a) as the 'medication revolution' (p. 83). As psychiatry became drug-based and clinical psychology colluded with the biologisation of the concept of the mind (see above under 'Biologisation of mind'), in general, society in both the USA and the UK, gradually adopted ways of thinking that had been set up by psychiatry and backed by clinical psychology. These particular ways of thinking about the human condition have been promoted by a form of the 'looping effect' described by Hacking (1995, 1999) whereby psychological and psychiatric categories and constructs that circulate in the wider world (popularised and spread in this case by the 'psy' disciplines and the pharmaceutical industry) have become internalised by individuals to shape their experiences and observations. Essentially, these 'modern' ways of thinking regard many human problems in living as indications of illness; according to such views, those problems are reflections of biological events in the brain and so are best dealt with by individualised medications and/or packages of therapy directed at altering emotional states, individual beliefs or interactions between human beings —or a combination of these. Good psychiatric and psychological practice today means the ability to make clear diagnoses (see above) and administer specific therapies, usually drug remedies but also interventions aimed at changing people's mental functioning, belief systems and cultural practices.

There has been considerable criticism in the USA, since the early part of the twenty-first century, both of the excessive use of medications and of the Kraepelian model of mental illness that the 'psy' disciplines work with. Reviewing three important books for the prestigious *New York Review of Books*, Marcia Angell (2011), former editor of the prestigious *New England Journal of Medicine*, came to three significant conclusions, namely that: (a) pharmaceutical companies 'that sell psychoactive drugs through various forms of marketing, both legal and illegal, have come to determine what constitutes a mental illness and how these disorders should be diagnosed and treated' (Angell 2011, p. 3); (b) it is now highly doubtful that 'mental illness is caused by a chemical imbalance in the brain' (2011, p. 3); and (c) there is now convincing evidence that psychoactive drugs are not just useless as specific therapies, but may actually cause harm. However, in spite of publicity in the public domain criticising drug therapies, these continue to dominate the field of psychiatric practice, both in the USA and the UK. And the view held by most clinicians and a significant part of the general population in Euro-America is that diagnostic labels reflect specific illnesses that have a basic biomedical causation, although the ways these illnesses are experienced may be determined by social and cultural factors.

References

Alexander, M. (2012). *The new Jim Crow: Mass incarceration in the age of colorblindness*. New York: The New Press.

Angell, M. (2011). The epidemic of mental illness: Why? *The New York Review of Books*, June 23, 2011.

Banton, M. (1987). *Racial theories*. Cambridge: Cambridge University Press.

Barzun, J. (1937). *Race: A study in superstition*. New York: Harcourt Brace and Company.

Barzun, J. (1965). *Race: A study in superstition* (2nd ed.). New York: Harper & Row.

Barzun, J. (2000). *From Dawn to Decadence 500 years of Western Cultural Life 1500 to the Present*. New York: HarperCollins.

Bean, R. B. (1906). Some racial peculiarities of the Negro brain. *American Journal of Anatomy, 5,* 353–415.
Bernal, M. (1987). *Black Athena. The Afroasiatic roots of classical civilisation* (Vol. 1). London: Free Association.
Bonilla-Silva, E. (2014). *Racism without racists. Color-blind racism and the persistence of racial inequality in America* (4th ed.). New York: Rowman and Littlefield.
Bright, T. (1586). *A treatise of melancholy.* London: Vautrolier.
Burton, R. (1621). *The anatomy of melancholy* (11th ed.). London: Hodson. (1806).
Carew, J. (1992). The end of Moorish enlightenment and the beginning of the Columbian era. *Race and Class, 33*(3), 3–27. Retrieved on March 20 from http://journals.sagepub.com/doi/pdf/10.1177/030639689203300302.
Clifford, T. (1984). *Tibetan Buddhist medicine and psychiatry: The diamond healing.* Samuel Weiser: York Beach Main.
Davies, P., & Gribben, J. (1991). *The matter myth: Towards 21st century science.* Harmondsworth: Viking Penguin Books.
Dols, M. W. (1992). *Majnūn: The Madman in Medieval Islamic Society.* D. E. Immisch (Ed.). Oxford: Clarendon Press.
Donald, A. (2001). The wal-marting of American psychiatry: An ethnography of psychiatric practice in the late twentieth century. *Culture, Medicine and Psychiatry, 25*(4), 427–439.
Eze, E. (Ed.). (1997). *Race and the Enlightenment: A Reader.* Cambridge MA: Blackwell.
Eze, E. C. (2001). *Achieving our humanity. The idea of the postracial future.* New York and London: Routledge.
Fernando, S. (1991). *Mental health, race and culture.* London: Macmillan in association with Mind Publications.
Fernando, S. (2002). *Mental health, race and culture* (2nd ed.). Basingstoke: Palgrave.
Fernando, S. (2010). *Mental health, race and culture* (3rd ed.). Basingstoke: Palgrave.
Fernando, S. (2014a). *Mental health worldwide: Culture, globalization and development.* Basingstoke: Palgrave Macmillan.
Fernando, S. (2014b). Globalization of psychiatry—A barrier to mental health development. *International Review of Psychiatry, 26*(5), 551–557.
Finocchiaro, M. A. (1992). *The Gallileo Affair. A documentary history.* Oakland CA: University of California Press.
Foucault, M. (2006). *History of madness* (Jean Khalfa, Ed., J. Murphy and J. Khalfa, Trans.). London: Routledge.
Fromm, E., Suzuki, D. T., & de Martino, R. (1960). *Zen Buddhism and psychoanalysis.* London: Allen & Unwin.

Fryer, P. (1984). *Staying power. The history of Black people in Britain*. London: Pluto Press.
Galton, F. (1904). Eugenics: Its definition, scope, and aims. *American Journal of Sociology, 10*(1), 1–25.
Goldberg, D. T. (1993). *Racist culture. Philosophy and the politics of meaning*. Oxford: Wiley.
Gordon, L. (2015). *What Fanon Said. A Philosophical introduction to his life and thought*. London: Hurst.
Graham, T. F. (1967). *Medieval minds mental health in the Middle Ages*. London: Allen & Unwin.
Hacking, I. (1995). The looping effects of human kinds. In D. Sperber, D. Premark, & A. J. Premark (Eds.), *Causal cognition: A multidisciplinary approach* (pp. 351–383). Oxford: Oxford University Press.
Hacking, I. (1999). *The social construction of what?*. Cambridge, MA and London: Harvard University Press.
Haig-Brown, C. (1988). *Resistance and renewal: Surviving the Indian residential school*. Vancouver: Arsenal Pulp Press.
Hall, R. (1996). *Empires of the monsoon. A history of the Indian Ocean and its invaders*. London: HarperCollins.
Ingleby, D. (Ed.). (1980). *Critical psychiatry. The politics of mental health*. New York: Pantheon Books.
Johnstone, L., & Dallos, R. (2014). *Formulation in psychology and psychotherapy. Making sense of people's problems* (2nd ed.). London: Routledge.
Jordon, W. D. (1968). *White over Black. American attitudes toward the Negro 1550-1812*. Chapel Hill: University of North Carolina Press.
Kamen, H. (2014). *The Spanish inquisition. A historical revision* (4th ed.). New Haven: Yale University Press.
Kapuściński, R. (2008). *The other* (A. Lloyd-Jones, Trans.). London: Verso.
Karenga, M. (1982). *Introduction to Black studies*. Los Angeles: Kawaida Publications.
Khalfa, J. (2006). Introduction. In J. Khalfa (Ed.), M. Foucault *History of Madness* (J. Murphy and J. Khalfa, Trans., pp. xiii–xxv). London: Routledge.
Koyré, A. (1970). Introduction. In E. Anscombe and P. J. Geach (Eds.), *Descartes Philosophical Writings* (pp. vii–xliv). London: Nelson University Paperbacks for Open University.
Kraepelin, E. (1896). *Psychiatrie* (5th ed.). Leipzig: Verlagvan Johann Ambrosius Barth.
Kraepelin, E. (1913). *Manic depressive insanity and paranoia* (Trans. of *Lehrbuch der Psychiatrie* R. M. Barclay, 8th Ed., Vols. 3 and 4). Edinburgh: Livingstone.
Kuhn, T. S. (1962). *The structure of scientific revolutions* (3rd ed.). Chicago: University of Chicago Press.

Linnaeus, C. (1758–9) *Systema Naturae per Regina Tria Naturae* (10th Ed.). Stockholm: Laurentius Saluis (cited by Fryer, 1984).

Losurdo, D. (2014). *Liberalism. A counter-history*. London: Verso.

Martín-Baró, I. (1994). Toward a liberation psychology In A. Aron and S. Corne (Eds.). *Writings for a Liberation Psychology* (pp. 17–37). Cambridge, MA: Harvard University Press.

McQueen, D. V. (1978). The history of science and medicine as theoretical sources for the comparative study of contemporary medical systems. *Social Science and Medicine, 12*, 69–74.

Menzies, G. (2008). *1421 the year a magnificent Chinese fleet sailed to Italy and ignited the Renaissance*. London: HarperCollins.

Morrison, T. (1993). *Playing in the dark: Whiteness and the literary imagination*. London: Pan Macmillan.

Murphy, G. (1938). *An Historical introduction to modern psychology*. London: Routledge and Kegan Paul.

Outram, D. (2005). *The enlightenment* (2nd ed.). Cambridge: Cambridge University Press.

Pick, D. (1989). *Faces of degeneration. A European disorder, c. 1848-c. 1918*. Cambridge: Cambridge University Press.

Porter, R. (2002). *Madness. A brief history*. Oxford: Oxford University Press.

Porter, R. (2004). *Flesh in the age of reason. The modern foundations of body and soul*. New York: Norton.

Ryle, G. (1990). *The concept of mind*. London: Penguin Books. (First published by Hitchingson, New York, 1949).

Searle, C. (1992). 'Unlearning Columbus: A review article' *Race and Class, 33*(3), 67–77. Retrieved on December 22, 2016 from http://journals.sagepub.com/doi/pdf/10.1177/030639689203300306.

Shorter, E. (1997). *A history of psychiatry from the era of the asylum to the age of Prozac*. New York: Wiley.

Smith, G. (2008). *A short history of secularism*. London: Cambridge University Press.

Smith, A. (Ed.). (2015). *Artists and empire: Facing Britain's imperial past*. London: Tate Publishing.

Stannard, D. E. (1992). *American Holocaust. The conquest of the new world*. New York: Oxford University Press.

Tharoor, Shashi. (2016). *An era of darkness: The British Empire in India*. New Delhi: Aleph Book Company.

Walvin, J. (1993). *Black Ivory a history of British Slavery*. London: Fontana Press.

Woodward, C. V. (1974). *The strange career of Jim Crow* (3rd Rev. Ed.). New York: Oxford University Press.

CHAPTER 3

Race Thinking and Racism Become the Norm

The Atlantic slave trade that decimated the black populations of West and central Africa was accepted by most white Europeans largely because of racism—the notion that non-white races were inferior to others. After the abolition of laws legitimising slavery around 1807, followed by its actual abolition in the Americas many years later, racism became a crucial ingredient of the colonialism imposed on Asia and Africa, a project made possible by the wealth created in Europe by the plunder of the American continent and by race-slavery (see Chap. 2).

3.1 Effects of Colonisation

India has been referred to as the 'jewel in the [British] crown'—a term coined by Disraeli (1872) when he was the Conservative Party's opposition leader in parliament (see Kebbel 2010)—and its plunder was the mainstay not only of the UK's wealth but also of the impoverishment of its people. Under the rule of Queen Victoria, India's industry was suppressed and the economy of the country changed from being an exporter of manufactured goods to a producer of raw materials and a market for British goods (Panikkar 1959; Moorhouse 1983; Cotterell 2010; Tharoor 2016). The initial respect for India and its civilisation among Europeans of the East India Company (Willinsky 1998; Dalrymple 2003) changed, under direct British rule, to racist attitudes of white superiority; this rule was imposed after a mutiny of *sepoys* (Indian soldiers under European command) in 1857 developed into a general uprising (the First War of Independence), which was suppressed by British troops with unparalleled savagery. From then on, British policy in India under direct rule by the British crown, as reflected in the activities of expatriates on the ground, was restructured to fit into the notion of 'The White Man's Burden', the phrase coming from a poem written in 1899 by Rudyard Kipling, which underpinned 'a theme throughout history [since the European Enlightenment] of the West and the Rest' (Easterly 2006, p. 19). The view grew up in the West—and was then exported universally as European power spread—that the countries of Asia and Africa had *always* been backward or underdeveloped until Europeans came along; even Karl Marx (1978) wrote in 1853 that India's 'social condition has remained unaltered since its remotest antiquity until its final decennium of the nineteenth century' (p. 656). From the late eighteenth century onwards, racism and colonialism reinforced each other; and non-white natives of the colonies were seen as backward people unable to rule themselves because they were racially inferior. William Wilberforce, a leader of the fight against the slave trade, considered that the conversion of India to Christianity was a cause *greater* than the abolition of slavery: 'He told the House of Commons in 1813, in the debate which preceded the new India Act [which ended the East India Company's monopoly of trade with India] that he saw the sub-continent as a place which would "exchange its dark and bloody superstition for the genial influence of Christian light and truth", the gods of the Hindus being "absolute monsters of lust, injustice, wickedness and cruelty. In short, their religion is one grand abomination"' (Moorhouse 1984, p. 69).

Finally it was Africa's turn to be exploited—this time not just for cheap labour (as during the Atlantic slave trade) but for its wealth, its minerals, gold and precious materials—a process that was accompanied by unprecedented atrocities and suffering being inflicted on its people. And, this time round, the exploitation affected not just sub-Saharan Africa but, eventually, nearly the whole continent, including North Africa, and comprising the Levant and the Maghreb, the areas that had been part of the great Islamic Empire of the Middle Ages (Robinson 1996). Colonisation was carried out piecemeal but soon became a competitive project leading to conflict between European powers. At the Berlin conference in 1884/1885, most of Africa was divided between the spheres of influence of the various European powers pursuing colonisation—a process otherwise known as *The Scramble for Africa* (Pakenham 1992). Colonisation of the continent was completed by about 1920 (Davidson 1984), except for Ethiopia, then called Abyssinia, which was subjugated by Italy's fascist government in

1936. The nations of Africa, which had been as distinct as those of Europe before the nineteenth century, were destroyed as entities as tribalism was encouraged in the interest of easier colonial rule, following the divide-and-rule model (Davidson 1984). African cultures were destroyed, and the colonies were underdeveloped and remodelled economically to provide raw materials and markets for European nations to exploit (Rodney 1988). The colonisation of Africa was more brutal than that of Asia, the worst atrocities having been committed in the Belgian Congo, a vast expanse of land held as his personal fiefdom by (Belgian) King Leopold II, a cousin of (British) Queen Victoria (see *King Leopold's Ghost* by Hochschild 2000); and in the German colony of German south-west Africa (covering mainly what is now Namibia)—see *The Kaiser's Holocaust* by Olusoga and Erichsen (2010). The whole colonisation project was underpinned and condoned by most European countries because of the solidity of the doctrine of racism was held in Europe with almost religious fervour. Hochschild (2000) writes: 'During the nineteenth-century European drive for possessions in Africa and Asia, [European] people justified colonialism in various ways, claiming that it Christianized the heathen or civilised the savage races or bought everyone the miraculous benefits of free trade. Now with Africa a new rationalization had emerged: smashing the "Arab" slave trade' (p. 38).

3.2 Power of Racism

The racism which spread all over the world (to varying degrees), and was particularly in evidence wherever Europeans managed to get to and establish a presence, must be seen in association with the invention of the 'white race' (see Allen 1997) and the notion of white supremacy. Chapter 2 describes how racism was constructed in the 'New World' of America and soon took a virulent form, justifying atrocities and the genocide of both indigenous peoples and people who were brought there from Africa by (mainly British) slave traffickers. Many of the immigrants from Europe already had some concept of whiteness and soon came to distinguish themselves from the 'Other' by means of this concept—by race. These Others were (a) native people of the Americas (mistakenly called 'Indians' because the Spanish sailors who first got to the American continent thought they had reached India); (b) enslaved Africans transported across the Atlantic, first by the Spanish and Portuguese but soon almost exclusively by the British; (c) indentured labourers from Ireland, where they had already been racialise by the English colonisation of their country; and Jews, who were traditionally seen in Europe as racially different to white people (see 'Exploration, colonialism, race-slavery' in Chap. 2 and 'Racialisation' in Chap. 5). In time both the Irish and Jews were absorbed into 'whiteness' in the USA and Canada (see *How the Irish Became White* by Ignatief 1995, and *How Jews Became White Folks and What That Says About Race in America* by Brodkin 2010), although newcomers in later years, such as Muslims from West Asia (the Middle East) and brown-skinned people from Asia, have tended to be viewed as outside the dominant white majority and are usually referred to as an ethnic group (see 'Islamophobia' in Chap. 8).

Slavery was abolished legally in the USA by the Thirteenth Amendment to the US Constitution: Abolition of Slavery (1865), which decreed that: 'Neither slavery not involuntary servitude, except as a punishment for crime whereof the party shall have been duly convicted, shall exist within the United States, or any place subject to their jurisdiction'. John Patterson (2017), reviewing for *The Guardian* a 2016 documentary film directed by Ava DuVernay, points out that the clause 'except as a punishment for crime whereof the party shall have been duly convicted' enabled southern states in need of cheap labour to 'shift the laws to make more criminals, then make them work for free as prisoners'. Patterson quotes 'prison reform activist and ex-con Glen E. Martin [saying] "Every time you give me liberty, the handcuffs come out right after"' (pp. 10–11). The film itself highlights the racial oppression that has continued in the USA from the days of slavery to the present day (2017). Significantly there were more black men in prison at a time when a black man (Barack Obama) was elected president of the USA than there were enslaved black people in 1850 (Alexander 2012) (see also 'The Obama years' in Chap. 8).

Human relationships are often structured by the relative power of interacting groups in society, however derived; and overt domination on a large scale of groups of people (as in the case of enslavement of Africans during the Atlantic slave trade and colonialism applied by Europeans to people outside its own continent) requires a philosophy to justify it, at least at a superficial level (see also the section 'How whiteness operates' in Chap. 7). That may well be the main reason for the notion of 'race' becoming so important in European culture. In other words, the doctrine of the racial inferiority of certain groups of people—racialised people—is necessary, both sociopolitically *and* psychologically, in order to maintain power relationships locally and geopolitically—to maintain the position of Europeans, 'white people' vis-à-vis other types of people—other 'races', 'cultures', ethnic groups, underdeveloped nations of the Third World and so on.

3.3 Distortions of History

Historiography, the study of written history or the writing of history, has been explored from various perspectives, the most influential being that of Georg Wilhelm Friedrich Hegel (1770–1831). In his book *Learning to Divide the World: Education at Empire's End*, John Willinsky (1998) discusses some of the basic features of Hegel's analysis of history. In a book transcribed from lectures delivered in Berlin during the 1820s, Hegel (2004) saw 'reason', the idea that one event follows another in a fashion that is reasonable, as the force governing history, and proposed that participation in history depends on people's knowledge of the 'Spirit of Freedom'. According to Hegel 'Orientals have not attained the knowledge that Spirit—Man *as such*—is free (p. 18, emphasis in original) so that China and India stand 'outside the World's History' (p. 116), being '*entirely* wanting in the essential consciousness of the idea of Freedom' (p. 71, emphasis in original). And in the case of Africa, '[w]hat we properly understand by Africa, is the Unhistorical, Undeveloped Spirit, still involved in the conditions of mere nature and which had to be presented here only as on the threshold of the World's History' (p. 99). For Willinsky (1998), 'what fascinates [...] is the degree to which Hegel's suppositions about history and the West's hold on history have become part of what we regard as a commonsense historical understanding' (p. 116)—and this is what drives the writing of history. In effect, the racist perceptions, and the institutional racism that pervade the Hegelian understanding of history structure what is accepted as 'true', what is correct history.

By the mid-nineteenth century the 'natural' superiority of Europe (populated by white people) became an article of faith; and the term 'primitive' was applied indiscriminately to non-white people all over the world (Worsley 1972). The histories and achievements of African, pre-Columbian American and Asian cultures were discredited. Although highly developed systems of social organisation had flourished in Africa for many centuries before the European invasion (Davidson 1974), the renowned historian Trevor-Roper, who wrote the authoritative *The Rise of Christian Europe* (Trevor-Roper 1966), argued in a talk on British television that African history south of Roman North Africa was non-existent before imperial (European) rule (Trevor-Roper 1963).

In his book *Black Athena* Martin Bernal (1987, pp. 29–30) shows how the 'paradigm of "races" that were intrinsically unequal in physical and mental endowment' was applied to the interpretation of history (see 'Exploration, colonialism, race-slavery' in Chap. 2 for a discussion of how China and Egypt were subjugated). The result was that the image of people associated with illustrious civilisations was changed by historians to suit the racist ideologies that informed the European culture of the nineteenth century. The image of China changed; it was now to be seen as 'a filthy country in which torture and corruption of all sorts flourished. With obscene irony, the Chinese were especially blamed for their consumption of opium' (1987, p. 238). According to Bernal, until the early nineteenth century, the conventional view had been that Greek culture had arisen as a result of colonisation, in around 1500 BC, by Egyptians and Phoenicians, who had civilised the native inhabitants; but the writing of history changed in the nineteenth century to cite a cultural invasion of Greece from the north. The racial position of the Egyptians according to the black–white classification system presented a problem for historians: in the Middle Ages and Renaissance, 'Egyptophile masons (impressed by their engineering prowess, for example in constructing the pyramids) tended to see them [Egyptians] as white... [but]... the Hellonomaniacs of the nineteenth century began to doubt their whiteness and to deny that Egyptians had been civilized' (Bernal 1987, p. 30). 'Both [Chinese and Egyptians] were flung into prehistory to serve as a solid and inert basis for the dynamic development of the superior races, the Aryans and the Semites'. Egypt was somehow presented as 'white' while Egyptians themselves were seen as 'black' (pp. 29–30).

According to Billig (1982), a 'dramatic indication of the way in which racial presuppositions were spreading in the nineteenth century is provided by the admiration of both Engels and Darwin for the biological writings of Ernst Haeckel... identified as one of the most important forerunners of Nazism' (pp. 69–70). Haeckel's *The History of Creation* (1876) was praised by Engels and his *The Riddle of the Universe* (Haeckel 1901) was similarly praised by Lenin. Both books argued for the inequality of races; the latter, developing notions of 'Aryan' superiority, claiming that Caucasians have 'from time immemorial been placed at the head of all races of men, as the most highly developed and perfect' (Haeckel 1876, p. 321). Darwin (1871) in *The Descent of Man* argued against the 'races of man' being 'distinct species' but clearly supported the view of white superiority: 'When civilized nations come into contact with barbarians the struggle is short, except where a deadly climate gives its aid to the native race' (p. 283). The application of Darwin's theory of evolution to social and psychological fields (by his cousin, Francis Galton) led to the eugenics movement, which in turn provided the racist dogma for the Jewish Holocaust and European fascism of the 1930s.

3.4 THE ARTS AND NINETEENTH-CENTURY SOCIOLOGY

The infiltration of racism into Western painting in the eighteenth and nineteenth centuries has been explored by Sander Gilman (1985, 1992). Gilman (1992) records that an African woman called Saartjie Baartman who was exhibited in Britain and France became known as the 'Hottentot Venus' (p. 178); and '[i]n the course of the nineteenth century, the female Hottentot [woman] comes to represent the black female,... [t]he antithesis of European sexual mores and beauty is embodied in the black, and the essential black, the lowest rung on the great chain of being, is the Hottentot' (pp. 172–177). Claire Pajaczkowska and Lola Young (1992) write how historical research by Catherine Hall (1988, 1991) at University College, London shows 'the [English] literature and imagery of the time [the nineteenth century] are replete with examples of fantasies of indolence and greed; of natives [meaning black people] waiting passively to be fed, without effort, by bountiful nature, fantasies of uncontrolled sexuality and fecundity, ... [all of which] bore no relation to the actual predicament of the indigenous peoples [of Europeans colonies], but formed what can now be recognized as the disowned, "split off" or disavowed aspects of the "independent" man's self identity' (Pajaczkowska and Young 1991, pp. 202–203). In much of European literature 'black' came to represent ignorance, aggression and violence; and black sexuality was seen as a threat to white people.

The sociology that developed in the nineteenth century reflected the division of the world into two groups, the civilised white people originating in Europe and the savage races native to Asia, Africa and America. Sociologists studied white people while social anthropologists, in alliance with physical anthropologists, studied primitive [*sic*] races. In tracing European myths of the Orient, Rana Kabbani (1986) has observed that nineteenth-century anthropology, being predominantly concerned with the hierarchical classification of race, was inextricably linked to the functioning of empire and imperialism. Edward Said (1978) writes how as Western powers pounded China into submission, 'Jesuits had opened up the new study of China, ... [and]... by the middle of the nineteenth century Orientalism was as vast a treasure-house of learning as one could imagine' (p. 51). Two types of people were constructed as inhabiting Western Asia (renamed the Middle East): 'the sacred and the profane (Jews and Christians in the first, the Muslims in the second)' (p. 64).The 'Mohammedan' was manufactured as 'the imitation of a Christian imitation of true religion... [and] Mohammed no longer roams the Eastern world as a threatening, immoral debauchee; he sits quietly on his (admittedly prominent) portion of the Orientalist stage' (pp. 64–66)—belonging for ever to Europe—but clearly not European. Western Asia is thus pulled into the Western world, Islam is insulted and (in a way) the colonial scene is set.

3.5 Nineteenth-Century Psychology and Psychiatry

During the colonial period, psychiatrists and psychologists, like other academics and clinical practitioners around them, had very clear-cut ideas on which races were civilised and which were not. A paper published in the mid-nineteenth century in the *Journal of Mental Science* (the forerunner of the *British Journal of Psychiatry*) by a former Physician Superintendent of Norfolk County Asylum who was working in Turkey, referred to that land as 'a country which forms the link between civilization and barbarism' (Foote 1858, p. 444); in the same journal, another British psychiatrist, Daniel H. Tuke (1858), writing on the association between civilisation and the generation of mental illness, denoted Eskimos, Chinese, Egyptians and American blacks as 'uncivilised' people, contrasting them with Europeans and American whites, who were referred to as 'civilised' people, but with a grudging reference to China as 'in some respects decidedly civilized' (1858, p. 108).

Although clinical psychology and psychiatry both embraced racism in their theories and clinical activities from their earliest inception, what stood out most in the early part of the twentieth century was the racism made explicit in the field of intelligence and cognition. The introduction of intelligence testing in the early twentieth century in the USA (later spreading to the UK and other parts of Europe), was stimulated by debates about 'Negro education' and eugenic fears about immigrants from Eastern European countries, especially Jewish people. According to Richards (2012): 'A large literature of mainly Southern-authored social commentary appeared as the twentieth century got under way, variously alarmist, meliorative, defensive, and desperate in tone' (p. 77). Psychologists researched alleged 'race differences' among children, thinking that this approach would offer a '"neutral", respectably scientific, route for addressing the intractable difficulties from a new angle—one that located the source of the problem safely at the individual psychological level, in the "Negro" psyche itself' (p. 77). Richards (2012) calls such a seemingly empirical, colour-blind approach, in a field that is beset by racist assumptions, 'empirical racism'—racism camouflaged by empiricism, a type of racism still evident today in the 'psy' disciplines. As European concerns around 'race' (often focusing on retaining racial purity) spread across the Atlantic, the preoccupation with the issue of 'Negro education' in North America was supplemented by eugenic concerns in what was seen as the 'melting pot' of immigrants from Europe. Thus, the legacy of slavery and eugenic fears aroused by immigration were combined and formed a backdrop to the 'race psychology' that developed in the USA in the early part of the twentieth century—just as a knowledge base was being forged for psychiatry on both sides of the Atlantic (see the section 'Race psychology' later in this chapter).

Meanwhile in Western Europe, especially Britain and Germany, the adaptation of Darwin's ideas by his colleague and half-cousin Francis Galton—as 'social Darwinism'—led to the eugenics movement. Eugenics established itself as a powerful dogma at the prestigious London School of Hygiene and Tropical Medicine (LSHTM) and equally prominent University College London in Gower Street. In the second of two papers on 'Hereditary Talent and Character', which is commonly held to mark the start of modern British psychology (Billig 1982, p. 72), Galton (1865), a renowned anthropologist and psychologist, claimed that European 'civilised races' alone possessed the 'instinct of continuous steady labour' while non-European 'savages' showed an innate 'wild untameable restlessness' and being 'incapable of progress... remain children in mind with passions of grown men' (pp. 325–326). Galton's views were regarded with much respect by the British establishment at the time and were certainly not regarded as eccentric or offensive. He received many awards during his career. He was made a fellow of the Royal Society in 1860, and was knighted shortly before he died (Brignell 2010). The main thrust of eugenics (accepted at the time as a scientific endeavour) was to identify 'inferior' races. Protégé of Galton and first holder of the Galton Chair of Eugenics at University College, Karl Pearson (1901) saw the extermination of such races as an inevitable part of the evolutionary process. Yet the racist eugenic movement (like most academic discourses) formed links across the

Atlantic into the wider English-speaking world. Work carried out on Ellis Island (where immigrants to the USA were held when they first arrived) by H. H. Goddard—America's leading eugenicist psychologist [at the time] according to Richards (2012, p. 78)—led to a scale for the estimation of 'mental defect' [*sic*] (Knox 1914).

3.6 Inherited Instincts

Instinct theory was a popular theme in the early twentieth century—and is still quoted sometimes—which was used to support the idea of inborn psychological differences between individuals and even to explain sources of hostility between groups of people. British academic psychologist William McDougall (1908) in *Introduction to Social Psychology*, a book that would be the standard textbook on social psychology for many years after it was first published, explored the relationships between emotions and instincts in great detail. In McDougall's category of 'instincts', which he presented as promoting tendencies in human beings to exhibit a variety of emotions such as feelings of sympathy or anger, he included also tendencies to exhibit certain behaviours such as aggressiveness or submission, describing the latter in dogs, which look down when submitting to their master. In a later publication which he wrote after emigrating to the USA, *Is America Safe For Democracy*, McDougall (1921) expands this line of thinking to claim that 'in the Negro the submissive impulse is strong' and that 'in the great strength of this instinct of submission, we have the main key to the history of the Negro race' (pp. 118–119). 'In *The Group Mind* (McDougall (1920)' made detailed analyses of how different races produced different 'group minds', made up of largely innate (instinctual) tendencies. Although he conceded that the differences can be modified by physical and social environment, he saw them as fixed for ever in the relationships of races to one another, and between people from 'widely different groups of peoples such as the Negro, the White, and the Yellow' (p. 111)—in other words, he believed that the differences between the traditional 'races' of European culture (see Chap. 2) would never change. In explaining what he meant by these racially fixed tendencies, McDougall went on to claim that the '[N]egro race wherever found does present certain specific mental peculiarities... [such as]... happy-go-lucky disposition, the unrestrained emotional violence and responsiveness; whether its representatives are found in tropic Africa, in the jungles of Papua, or in the highly civilised conditions of American cities' (p. 111); and made similar claims about other racial groups.

3.7 Race Psychology

According to Richards (2012), Herbert Spencer (of eugenics fame) designated race psychology as the 'science of the evolution of mind in animals and man' (p. 79). US Army data on cognitive test results gathered during the 1914–1918 war led to a discussion of the reasons for racial differences in scores on intelligence tests (IQ tests). In an influential book *The Measurement of Intelligence*, Lewis Terman (1916), the eminent psychologist from Stanford University, claimed that Negroes, Spanish-Indians and Mexicans were of low intelligence because of their race. This was the beginning of what Thomas and Sillen (1972) call the 'racist IQ movement' (p. 34) which gathered momentum after the WWI, with support from army data published by the National Academy of Sciences under the editorship of Robert Yerkes (1921). Although beginning with the study of intelligence, the body of knowledge included under the rubric of 'race psychology' reflected the intense interest of Western psychology in analysing in a framework of racial difference anything to do with (a) the functioning of the human mind, or anything seen by psychologists as related to it—memory, perception, instincts, and so on—and later, (b) with psychological aspects of matters discussed in the allied disciplines of anthropology and sociology.

Race psychology encompassed a range of subjects all analysed on the basis of an inherent 'difference' between individuals attributed to 'race', as that concept had developed in Western culture (Chap. 2). Richards (2012) has collected together and tabulated (pp. 80–81) some of the vast amount of literature on race psychology in the English language published between 1909 and 1940—more or less between the two world wars, nearly all of which emanated from white psychologists in USA, supplemented by the work of Japanese and Chinese psychologists. The work included studies of alleged innate characteristics of human beings, which informed educational policies; eugenic studies that informed immigration policies about alleged proclivity to alleged criminal tendencies and psychopathology and psychological instability; so-called scientific mapping of race differences in physiology; and research into the alleged complexities of 'race-mixing'. Indeed in the first two decades of the twentieth century, *psychology was about 'race'* (reflecting the European racist thinking that had been set down during slavery and colonialism), and was dominated by eugenic thinking mixed up with aspects of the study of social class and socio-economic efficiency. This pre-occupation of academic psychology with 'race' is largely limited to discourse in Europe and North America and certainly not a feature of non-Western psychologies of Asia, Africa and so on. This focus in Europe may have affected the politics of both the left and the right: In the 1930s, British Fabians, including George Bernard Shaw and William Beveridge, mastermind of the welfare state, in their aspirations for democratic socialism at home supported eugenic policies (Freedland, 2012), and the 'German cultural climate of the 1930s favoured racist assumptions as "respectable" [science]' (Billig, 79, p. 5). In the USA, the Rockefeller Foundation provided the finance for one of the most controversial of all political developments associated with eugenics: the work of (psychiatrist) Ernst Rüdin.

In 1918, Kraepelin set up the German Psychiatric Research Institute in Munich with his pupil, Ernst Rüdin, as the head of its Genealogical Department (Weindling 1989). The institute stressed that its aim was to protect the public from dangerous and burdensome mentally ill people, and much of its early work consisted of establishing a data bank of people seen in these terms. Its main research thrust was to investigate the genetic patterns of what were assumed to be inherited diseases, including schizophrenia, mainly by studying histories of twins (Weindling 1989). A major result of the work by this institute was the sterilisation campaigns of the 1930s and finally the actual medical killing of people diagnosed by psychiatrists as incurably schizophrenic. Rüdin's collaborators were Eliot Slater and Franz Kallman—the former migrated to the UK where he held a senior post at the Institute of Psychiatry (IOP) and the latter, who had Jewish ancestry, migrated to the USA. None of these psychiatrists faced the post-war Nuremburg courts. Undoubtedly, the race psychology of the early twentieth century left a deep mark, if it did not in fact determine the future of (Western) psychology and psychiatry as we see it today—most notably in the emphasis in the 'psy' disciplines on inheritance as causative of psychological pathologies and mental illnesses.

3.8 Mental Pathology and the Construction of Race-Linked Illnesses

In the mid-nineteenth century, John Langdon Down, a distinguished British doctor who was superintendent of an institution for so-called 'idiots' in South London, wrote a paper constructing what he called an ethnic classification of these 'idiots' (Down 1866). In this he claimed that so-called 'idiots and 'imbeciles' of 'European descent' (that is white parentage) showed physical characteristics of Ethiopian, Malay, (indigenous) American and Mongolian racial types, being racial throwbacks to 'one of the great divisions of the human family', namely races; and that 'a very large number of [so-called] congenital idiots are typical Mongols' (1866, p. 16). It should be noted that this was a time when people with learning difficulties were highly stigmatised and were seen as suffering specific illnesses (or 'conditions') that were inherited. At about the same time, Charles Darwin's classic *The Expression of the Emotions in Man and Animals*, (Darwin 1872) devoted a whole chapter to the study of human facial expression. Blushing and conscience were thought to be related and the debate that ensued about the capacity of 'Negroes' to blush was 'not so much a physiological one, as one about moral development' (Skultans 1979, p. 63). Although Darwin argued against the 'races of man' being distinct species, he saw the domination of white races as a natural development in evolution. In fact, Darwin's groundbreaking *On the Origin of Species by Means of Natural Selection* (1901) was subtitled *Or the Preservation of Favoured Races in the Struggle for Life*.

A classic textbook on adolescence written by Stanley Hall (1904), the first president of the American Psychological Association, founder of the *American Journal of Psychology* and perhaps the originator of modern (Western) clinical psychology, contained the first description (in Western culture) in a book on psychology, *by a psychologist,* of what would now be called 'racialised people' (see 'Racialisation' in Chap. 5), referred to (in the book) as Indians, Africans and North American 'Aborigines', and likened by Hall to immature children who 'live a life of feeling, emotion and impulse' (Stanley Hall 1904, p. 649). Hall's description of racialised people was in keeping with that of psychiatrist Kraepelin (1921) who saw Javanese people as being akin to *immature* European youth (p. 171, emphasis in original), and was also in line with the myths about these peoples that had grown up in Europe and North America as slavery and colonialism thrived (see 'Exploration, colonialism, race-slavery' and other sections in Chap. 2).

In the clinical field, the social construction of mental illness is shown up dramatically in the decision of the American Psychiatric Association (APA) in 1973 that homosexuality should cease to be classed as an 'illness' (Bayer 1981)—a decision taken by a majority vote of the APA. Similarly, racist considerations are evident in the construction of categories of mental illness reported in the USA at the time of slavery—in particular, conditions described by Cartwright (1851), the pre-eminent writer on the topic of diseases among enslaved people in the USA, as peculiar to black people.

The most (in)famous illness described by Cartwright is '*Drapetomania* or the disease causing slaves to run away' (p. 318). After attributing the condition to 'treating them [black people] as equal' or frightening them by cruelty, Cartwright advocated a mixture of 'care, kindness, attention and humanity', with punishment 'if any one or more of them, at any time, are inclined to raise their heads to a level with their master or overseer… until they fall into that submissive state which was intended for them to occupy' (p. 320). Another alleged disease among enslaved black people, *Dysaesthesia Aethiopis*, was described by Cartwright as 'affecting both mind and body… [with] insensibility of the skin… and hebetude of intellectual faculties… much more prevalent among free slaves [meaning people freed from being enslaved] living in clusters by themselves than among slaves [meaning enslaved people] in our plantations' (p. 320). He claimed that people affected by this disease 'break, waste and destroy everything they handle… slight their work … and raise disturbances'; and suggested a regime of 'treatment' based on 'sound physiological principles', which consisted of oiling the skin and 'to slap the oil in with a broad leather strap, … hard work in the open air … [and] good wholesome food' (pp. 322–323).

Almost into the twentieth century, Babcock (1895), a psychiatrist from South Carolina, was using pro-slavery arguments to develop the theme that Africans were inherently incapable of coping with civilised life. Babcock juxtaposed the idea that mental disease was 'almost unknown among savage tribes of Africa' with alleged observations in the USA on the 'increase of insanity [among African Americans] since emancipation', which he considered was due to the deleterious effect of freedom on 'sluggish and uncultivated brains' and 'the removal [during emancipation] of all healthy restraints'; he forecast 'a constant accumulation of [black] lunatics' in the years to come (pp. 423–425). The fear of black lunatics is one that seems to have haunted white Euro-Americans from the very early days of black–white interaction and is reflected in discussions about 'civilisation' and mental disorder (see the section 'Alleged mentality of black people' in Chap. 4; and 'Racialisation of the schizophrenia diagnosis' in Chap. 5).

Carl Jung, one of the most eminent of Western psychologists, fancied himself as a specialist on black people since he had actually visited Asia and Africa. When he visited the USA in the 1920s, Jung (1930) felt dissatisfied at being unable to 'size them up', referring to the white population; he could not, at first, understand 'how the Americans descending from European stock have arrived at their striking peculiarities' (pp. 193–195). He focused on 'the Negro' as the cause. In postulating a psychological danger to white people of living in close proximity to black people, Jung deduced the theory of 'racial infection—'a very serious mental and moral problem wherever a primitive race outnumbers the white man' (pp. 193–196):

> Now what is more contagious than to live side by side with a rather primitive people? Go to Africa and see what happens. When the effect is so very obvious that you stumble over it, then you call it 'going black'... The inferior man exercises a tremendous pull upon civilized beings who are forced to live with him, because he fascinates the inferior layers of our psyche, which has lived through untold ages of similar conditions. (p. 196)

It would seem that the 'racial infection' Jung claimed to have identified was some sort of psychological transference of behaviour patterns, because the 'peculiarities' that he felt concerned about in (white) Americans were largely about the way they walked and spoke—for example he observed that (white) Americans seemed to sway their hips like Negroes (Jung 1921). Quoting an observation that the experience of a 'savage' during a dream was just as real to him as what he saw when he was awake, stated: 'What I myself have seen of the psychology of the Negro completely endorses these findings' (p. 30). Clearly, Jung identified the modern African as 'primitive' in every sense of the word; and then went on to see all non-Europeans—racialised people—in similar terms, as people who could not separate themselves out as individuals, in whose minds object and subject were not differentiated and whose feelings were concreistic—the antithesis of abstraction.

Considering his reputation and the fact that his work is quoted often by psychologists, Jung's racist legacy to psychology has evoked very little comment from clinical psychologists on both sides of the Atlantic. British psychologist and group analyst Farhad Dalal (1988) is almost unique among psychologists in pointing out how deeply racist many of Jung's theories (that are directly applicable in clinical work) are. In his study of the racism of Jung, illustrated by many quotes from Jung's work, Dalal (1988) states: 'In the "growth movement" one hears constant accolades on Jung. He is revered for several things. He is supposed to be the father of Transpersonal Psychology; the man who unified the human race through his concept of the collective unconscious, and then connected the human race to the greater cosmos; it is said that he is the great equaliser and the great unifier; that his philosophy is that of balance and humility. And it is true that he has done these things, but only in part and at a cost, the cost being not only a retention but also a reinforcement of the *status quo* and the iniquities contained therein' (p. 263). According to Dalal, Jung's model for the mind of the infant was very similar in many ways to that of (what Jung called) 'primitive' humans, a description that clearly referred to anyone who was not white. In fact, 'Jung uses the word "primitive" in two senses: (1) as the prehistoric human, and (2) the modern black.... To Jung they are all one' (p. 264). Dalal expresses astonishment that Jung, being a 'cultured' and educated man who had studied Sanskrit and Pali books, had 'pillaged them when he wanted to substantiate some of his more mystical notions'; he (Jung) *rubbishes Indian culture and emotional and mental faculties of the oriental and on the other hand he uses their spiritual concepts freely, calls then advanced, and at times claims them as his own'* (p. 269, emphasis in original). Dalal concludes that Jung equated the white unconscious with the black conscious, and then assumed that what he could discern of his own unconscious life represented the symbolism used by black people.

Jung's work stands out as a vivid example of the ways in which racism became woven into psychological thinking in the mid-twentieth century, and provides some insight into how invidious and powerful racist thinking can be in psychology. Dalal sees Jung's racist work as an abnormal event due to a personal quirk or psychopathology. A more realistic approach is to accept that racism is inevitable and normal in any Western psychological theory that addresses any aspect of culture or race unless specific efforts are made to exclude it. Jung is one of the very few Western psychologists who attempted to devise theories incorporating race and culture and he may not have recognised the extent to which his thinking was fashioned by racist notions—a criticism that may be applied to most specialists, even those currently active, when they formulate new ways of working and (for example) write reports such as *Understanding Psychosis and Schizophrenia*, considered in the section 'Racism of a psychology report' section in Chap. 8. The intricate ways in which this underlying racism has changed in the years since the end of WWII, especially from the 1970s onwards, is discussed in the following chapters.

REFERENCES

Alexander, M. (2012). *The new Jim Crow. Mass incarceration in the age of colorblindness*. New York: The New Press.
Allen, T. W. (1997). *The invention of the white race. The origin of racial oppression in Anglo-America*. London: Verso.
Babcock, J. W. (1895). The colored insane. *Alienist and Neurologist, 16*, 423–447.

Bayer, R. (1981). *Homosexuality and American Psychiatry: The politics of diagnosis*. New York: Basic Books.
Bernal, M. (1987). *Black Athena. The Afroasiatic roots of classical civilisation* (Vol. 1). London: Free Association.
Billig, M. (1982). *Ideology and social psychology*. Oxford: Blackwell.
Brignell, V. (2010). The eugenic movement Britain wants to forget, New Statesman 9 December 2010 Retrieved on June 14, 2016 from http://www.newstatesman.com/society/2010/12/british-eugenics-disabled.
Brodkin, K. (2010). *How the Jews became white folks ad what that says about race in America*. New Brunswick NJ: Rutgers University Press.
Cartwright, S. A. (1851). Report on the diseases and physical peculiarities of the Negro Race, *New Orleans Medical and Surgical Journal*, May, 1851, pp. 691–715; reprinted in A. C. Caplan, H. T. Engelhardt & J. J. McCartney (Eds.), *Concepts of health and disease* (Reading, Mass: Addison-Wesley) 1981.
Cotterell, A. (2010). *Western power in Asia. Its slow rose and swift fall 1415–1999*. Singapore: John Wiley.
Dalal, F. (1988). Jung: a Racist. *British Journal of Psychotherapy*, 4(3), 263–279.
Dalrymple, W. (2003). *White Mughals. Love and betrayal in eighteenth-century India*. London: Harper Collins.
Darwin, C. (1871). *The descent of man and selection in relation to sex* (Vol. 1). London: John Murray.
Darwin, C. (1872). *The expression of the emotions in man and animals*. New York: Appleton. (Reprinted 1965, London: University of Chicago Press).
Davidson, B. (1974). *Africa in history*. London: Paladin Books.
Davidson, B. (1984). *The story of Africa*. London: Mitchell Beazley.
Disraeli, B. (1872). The Maintenance of Empire speech (Reproduced in *Selected speeches of the Earl of Beaconfield*, pp. 529–534. Vol. II, by T. E. Kennel, Ed., 2010, London: Nabu Press). Retrieved on February 18, 2017 from http://www.ccis.edu/faculty/dskarr/discussions%20and%20readings/primary%20sources/disraeli,%20speech%201872.htm.
Down, J. L. M. (1866). Observations on an ethnic classification of idiots, *Lectures and reports from the London hospital for 1866*. (Reprinted *The Origins of Modern Psychiatry*, pp. 15–18, by C. Thompson, Ed., 1987, Chichester: Wiley.
Easterly, W. (2006). *The white man's burden. Why the West's efforts to aid the rest have done so much Ill and so little good*. Oxford: Oxford University Press.
Foote, R. F. (1858). The condition of the insane and the treatment of nervous diseases in Turkey. *The British Journal of Psychiatry*, 4, 444–450.
Freedland, J. (2012). Eugenics: the skeleton that rattles loudest in the left's closet. *The Guardian*, 17 February, 2012 Retrieved on 15 March 2017 from https://www.theguardian.com/commentisfree/2012/feb/17/eugenics-skeleton-rattles-loudest-closet-left

Galton, F. (1865). 'Hereditary talent and character' part 2. *Macmillan's Magazine,* 12, 318–327.
Gilman, S. (1985). *Difference and Pathology, Stereotypes of Sexuality, Race and Madness.* Ithaca: Cornell University Press.
Gilman, S. L. (1992). Black Bodies White Bodies. Toward an iconography of female sexuality in the late nineteenth-century art, medicine and literature, In J Donald & A. Rattansi (Eds.), *'Race', Culture and Difference* (pp. 171–197). London: Sage.
Haeckel, E. (1876). *The history of creation,* (Vol. 1, E. R. Lankester, trans.). London: Henry S. King.
Haeckel, E. (1901). *The riddle of the universe at the close of the nineteenth century.* London: Watt's. cited by Billig, 1982.
Hall, C. (1988). The economy of intellectual prestige: Thomas Carlyle, John Stuart Mill and the case of Governor Eyre. *Cultural Critique,* 12, Spring cited by Pajaczowska and Young, 1992.
Hall, C. (1991). Missing stories: gender and ethnicity in England in the 1830s and 40s. In L. Grossberg, C. Nelson & P. Treitler (Eds.), *Cultural Studies Now and in the Future.* London: Routledge.
Hall, G. S. (1904). *Adolescence its psychology and its relations to physiology, anthropology, sociology, sex, crime, religion and education* (Vol. II). New York: D. Appleton.
Hegel, G. W. F. (2004). *The Philosophy of History* (J. Sibree, trans.). New York: Dover.
Hochschild, A. (2000). *King Leopold's ghost. A story of greed, terror and heroism in colonial Africa.* Basingstoke: Macmillan.
Ignatief, N. (1995). *How the Irish Became White.* New York and London: Routledge.
Jung, C. G. (1921). *Psychologische Typen.* Zurich: Rascher Verlag. *The psychology of individuation* (H. G. Baynes, trans.). London: Kegan Paul.
Jung, C. G. (1930). Your Negroid and Indian behaviour. *Forum, 83*(4), 193–199.
Kabbani, R. (1986). *Europe's myths of orient. Devise and rule.* London: MacMillan.
Knox, H. G. (1914). A scale based on the work of Ellis Island for estimating mental defect. *Journal of American Medical Association,* 62, 741–747.
Kraepelin, E. (1921). *Manic Depressive Insanity and Paranoia* (R. M. Barclay & G. M. Robertson Ed. and trans.). Edinburgh: Livingstone.
Marx, K. (1978). The British rule in India. In R. C. Tucker (Ed.), *The marx-engels reader* (2nd edn., pp. 653–658). New York: Norton.
McDougal, W. (1908). *Introduction to Social Psychology.* London: Methuen. Also available on July 20, 2016 at: https://archive.org/details/introductiontoso 020342mbp.

McDougal, W. (1920). *The group mind. A sketch of the principles of collective psychology with some attempt to apply them to the interpretation of national life and character.* Cambridge: Cambridge University Press.

McDougall, W. (1921). *Is America safe for democracy?* New York: Scribner's Sons. Retrieved on July 20, 2016 from https://archive.org/details/gb22UB1-sp-AwC.

Moorhouse, G. (1983). *India Britannica.* London: Harvill Press.

Olusoga, D., & Erichsen, C. W. (2010). *The Kaiser's Holocaust. Germany's forgotten genocide.* London: Faber.

Pajaczowska, C. & Young, L. (1992). Racism, Representation, Psychoanalysis. In J. Donald & A. Rattansi (Eds.), *'Race' Culture and Difference* (pp. 198–219). London, Newbury Park, California and New Delhi: Sage.

Pakenham, T. (1992). *The scramble for Africa.* London: Abacus.

Panikkar, K. M. (1959). *Asia and Western Dominance. A comprehensive study of the European impact on Asia, from Vasco da Gama to the mid-twentieth century.* London: Collier books. Originally published by Allen and Unwin, London 1959.

Patterson, J. (2017). Ava DuVernay on the legacy of slavery: "The sad truth is that some minds will not be changed", *The Guardian G2* 7 February 2017, 10–11. Retrieved on February 10, 2017 from https://www.theguardian.com/film/2017/feb/06/ava-duvernay-legacy-slavery-selma-oscars-13th-trump-era-america-racist-past-award.

Pearson, K. (1901). *National life from the standpoint of science.* London: Adam and Charles Black. Cited by Fryer, 1984.

Richards, G. (2012). *'Race', racism and psychology. Towards a reflexive history* (2nd ed.). London: Routledge.

Robinson, F. (Ed.). (1996). *The Cambridge illustrated history of the Islamic world.* Cambridge: Cambridge University Press.

Rodney, W. (1988). *How Europe underdeveloped Africa.* London: Bogle L'Overture Publications.

Said, E. (1978). *Orientalism.* New York: Random House.

Skultans, V. (1979). *English madness. Ideas on insanity 1580–1890.* London: Routledge and Kegan Paul.

Terman, L. M. (1916). *The measurement of intelligence.* Boston: Houghton.

Tharoor, S. (2016). *An Era of Darkness: The British Empire in India.* New Delhi: Aleph Book Company.

Thomas, A., & Sillen, S. (1972). *Racism and Psychiatry.* New York: Brunner/Mazel.

Trevor-Roper, H. (1963). The rise of christian Europe. *The Listener, 70*(1809), 871–875.

Trevor-Roper, H. (1966). *The rise of christian Europe* (2nd ed.). London: Thames and Hudson.

Tuke, D. H. (1858). Does civilization favour the generation of mental disease? *Journal of Mental Science, 4,* 94–110.

Weindling, P. (1989). *Health, race and German politics between national unification and nazism.* Cambridge: Cambridge University Press.

Willinsky, J. (1998). *Learning to divide the world. Education at empire's end.* Minneapolis: University of Minnesota Press.

Worsley, P. (1972). Colonialism and categories. In P. Baxter & B. Sansom (Eds.), *Race and social difference* (pp. 98–101). Harmondsworth: Penguin.

Yerkes, R. M. (Ed.). (1921). *Psychological examining in the United States Army. Memoirs of the National Academy of Sciences* (Vol. 15). Washington: Government Printing Office.

CHAPTER 4

New Racisms Appear in the 1960s

Chapter 2 traced the origins of the word 'race' in European culture and the way that different races were placed along a hierarchy (of races), denoted predominantly by differences in skin-colour, which was developed in Western Europe and later extended across the Atlantic to North America. Chapter 3 discussed how racism became incorporated into European culture—became a 'European value'. This chapter addresses the changes that occurred fifteen or more years after the end of the Second World War (WWII) in 1945/6, a period which saw the rise of the civil rights movement in the USA and the collapse of European empires in Asia and Africa, and when new forms of racism—or new racisms—emerged on both sides of the Atlantic.

4.1 Transformations After WWII

The end of WWII in 1945–1946 was a watershed moment for the white people of Europe and America. It saw an acceleration in the migration (which had begun in the early part of the twentieth century) of African Americans out of the southern states of USA to the urban Northeast, Midwest and West (Gibson and Jung 2012); a weakening of the system of legalised racial discrimination in the southern states of the USA because of the demands of industrialisation; and, finally the civil rights movement of the 1960s and 1970s. The result of all of this was that massive changes had taken root by the mid-1970s in terms of race scene and in the group power relations associated with it across the USA. Meanwhile it was evident to European powers ('old Europe'), that there was little support from the two major power blocks (then USA and Soviet Union) for continuing race-based colonialism—and so, once-powerful colonial empires linked to skin colour/race began to crumble. By 1959, the world as white Europeans had known it was changing —the French, driven out of their Asian possessions, were facing defeat in Algeria; the Belgians were moving quickly towards decolonisation; South Africa was threatened by the prospect of race war (averted by the statesmanship of Nelson Mandela); and Britain and France suffered from the Suez debacle in 1956 when the USA and the Soviet Union forced them both to pull out of the Egyptian territory they occupied in colonial style.

Following the creation of the United Nations (UN), the Western powers (or at least the victors in WWII), still dominated the geopolitical scene but were seemingly reconciled to leaving behind the era of colonialism that had been based on racist persecution and the exploitation of non-white populations, and reached towards a reappraisal of world relations, and what the world should be like in the future. But this thinking and planning (both at the UN and generally across the world) was not informed by a proper reckoning on how the West and the Rest had related to each other in the previous 'Dark Ages' (see beginning of Chap. 2): it was not the sort of reckoning that was attempted after WWII by means of the Nuremburg war crimes trials and which led to the retributive work of the democratic government that emerged in West Germany (Jansen and Saathoff 2009); nor was it like the Truth and Reconciliation Commission in South Africa, which was created after the abolition of apartheid in South Africa in 1994 (Moon 2009); nor did it resemble the reckoning hinted at in the title of Caroline Elkins' Pulitzer Prize-winning book *Imperial Reckoning*, which exposed 'the untold story of Britain's Gulag in Kenya'. Although lessons had apparently been learned about totalitarianism and anti-Semitism (see for example, books by Arendt 1958, 1970), there was no *reckoning* with the racist past—in the sense of 'an instance of working out consequences or retribution for one's actions' (*Concise Oxford English Dictionary*, edited by Soanes and Stevenson 2008, p. 1201)—that addressed the 450 years or so of skin-colour racism, the associated enslavement and transportation of people and the atrocities enacted in the continents of Asia, Australasia, Africa and America. There has not yet (in 2017) been a proper opportunity for the West—the people of the West supported by their governments—to learn about and take to heart the lessons of racism, not to speak of an opportunity for them to try to make amends for what had been done in their name. Inevitably, what happened after the end of colonialism was that policies towards the ex-colonial countries were governed by the attitudes towards the racial 'Other' that had been fashioned *during* the periods of colonialism and the Atlantic slave trade. This failure to learn lessons was to come back to haunt Europe and the extension of its culture in North America, as chickens came home to roost in the twenty-first century with the rise of terrorism and internal conflict in Europe (see Chaps. 8 and 9).

The UN, formed after the end of WWII, developed a body, the World Health Organization (WHO), to help develop health services in ex-colonies and other parts of the world that were industrially underdeveloped (still no admission of the reasons for this underdevelopment), and hence resource-poor. In 1950, UNESCO, the United Nations Educational, Scientific and Cultural Organization, made its well-known declaration *The Race Question* (UNESCO 1950) rejecting the notion of a hierarchy of races and affirming the lack of scientific evidence of innate *racially determined* differences in the human race. And in 1965, the UN (1965) adopted the International Convention on the Elimination of All Forms of Racial Discrimination (ICERD), with the aims of: affirming 'human rights and fundamental freedoms for all, without distinction as to race, sex, language or religion'; bringing colonialism to an end; and 'speedily eliminating racial discrimination throughout the world' (p. 1). William Easterley (2006), former economist at the World Bank and Professor of Economics at New York University, commenting on changes (after the end of WWII) in attitudes in the West towards people in ex-colonial countries, writes: 'There was a genuine change of heart [in the West] away from racism and towards respect for equality, but a paternalistic and coercive strain survived' (pp. 20–21). I often experienced this apparent paternalism both while at university and in my professional life—indeed it was somewhat comforting and certainly preferable to overt racism, like being spat at or abused. And like other black people I soon realised that it was no more than a cover for racism (see examples of personal experience illustrating this in the 'Racism in a government body' subsection in Chap. 6).

The ex-colonial powers coped with the fall of white-dominated empires in different ways. For example, France attempted to hold on to its possessions for several years and even resurrected colonial wars, while the UK let go of its colonies as quickly as possible—perhaps too quickly and with inadequate preparation. Both policies caused wars: in the case of the French, in Vietnam and Cambodia (which, with Laos formed French Indochina in colonial times) and in Algeria; and in the case of the UK, the massive upheaval and suffering resulting from the division of India into two states (Zakaria 2015). But the white-dominated West as a whole was faced with a major shift in thinking vis-à-vis 'race' and its attitude to the racial-Other.

In the UK after WWII, there was considerable immigration of black and brown-skinned people who became British citizens, welcomed by successive British governments for supplying the workforce necessary for the recovery of industrial capacity and for the development of the welfare state (especially the new National Health Service, the NHS). Similarly, North America too had an influx of new immigrants, white people from Europe escaping unstable social conditions following WWII and black and brown-skinned people leaving the collapsing European colonies in Asia and Africa. This post-war immigration led to racial tensions (adding to those already there in the USA from the days of race-slavery and Jim Crow) which erupted from time to time into violence, leading to the US civil rights movement of the 1960s.

In the UK, there were debates about immigration in the British parliament, often with a racist tinge in that the concern seemed to be about black people coming into the country, rather than immigration as such. Then in 1958 there were skirmishes between black residents of Notting Hill (London) and the (white) police; and similar events in Nottingham (Miles 1984; Pilkington 1988). And then in 1968 a prominent Tory Member of Parliament, Enoch Powell, delivered a speech (reproduced in the *Telegraph*, 2007) warning of dire consequences of so-called 'coloured immigration'—a speech that became known as the 'Rivers of Blood' speech because Powell quoted Roman poet Virgil's foreboding, 'I seem to see the River Tiber foaming with much blood'. Powell's speech attracted much support from white people from many parts of the UK although Powell himself was sacked from the Conservative government for his racist speech. In fact, Powell's speech remained as a 'toxic cloud floating above all political debate on race relations [and immigration] for many years' (Mansoor 2008) (see utterances by historian David Starkey in 2016, referred to in 'Civil unrest' in Chap. 8).

From the mid-1960s onwards, the UK government introduced increasingly stringent rules to limit immigration from Asia, Africa and the Caribbean as well as 'race relations legislation' aimed at limiting racial discrimination. However, racial problems intensified and finally, the Labour government brought in the Race Relations Act (1976) (see below), giving a message—an important one at that time—to the general public that (at least overt) expression of racism was unacceptable, and no longer a British value as it had been for hundreds of years before then. This was an important turning point in the UK. From the late 1970s onwards, there was a diminution of street attacks and abuse, and much less overt discrimination, but racism was expressed and acted out in subtle forms (as well as the more obvious ones), often recognised only in terms of concepts such as institutional racism (referred to earlier).

In the USA, with the fall of state-sanctioned segregation resulting from the civil rights movement of the 1960s, African Americans achieved full voting rights and access to other civil rights with no racial barriers; and there was a gradual diminution of explicit discrimination in public places and overt racist practices. Moreover, as in Europe, explicit racism became unacceptable in many part of the country. What is clear is that overt expression of racist sentiments may have lessened in the USA (as in the UK) in the late 1980s and 1990s, but racial inequalities in many fields, including the field of mental health, persisted in the shape of institutional racism. Also, what is still striking in the USA (compared to UK) is that race-based segregation, sometimes partial but in many places almost complete, has persisted as a result of (apparent) personal preference and sometimes social pressures (the two being connected). When I visited Chicago in the 1990s, I was surprised that different parts of the city were clearly inhabited predominantly by specific racial groups—black, Chinese and so on. I heard from the black psychiatrist who showed me around how the local council managed to keep the black (African–American) areas poor and deprived of essential services through various political and legal manipulations and abuse of power—while on the surface, everyone was supposed to be equal.

The parlous situation in terms of racial discrimination in the USA today (2017) could be summarised with a list of the following facts: 'Inner city minority schools [catering predominantly for black students] in sharp contrast to white suburban schools, lack decent buildings, are over-crowded, have outdated equipment—if they have any equipment at all—do not have enough textbooks for their students, lack library resources, are technologically behind, and pay their teaching and administrative staff less [and so on]. … One in three black males born today [in the USA] can expect to spend some portion of his life behind bars, and Latinos have seen a 43% rise in their incarceration rates since 1990. … Race differences exist at nearly every stage of the juvenile justice process; black youth suffer racial profiling, higher rates of arrest, detention and court referral, are charged with more serious offences, and more likely to be placed in large public correctional facilities in contrast to small private group homes, foster homes and drug and alcohol treatment centers' (Bonilla-Silva 2014, pp. 35–44). All this amounts to a high level of institutional racism, far worse than that in the UK.

During the decades following the end of WWII, the old-fashioned overt form of racism based on a narrow biological view of 'race' and unashamed expression of racial prejudice dwindled in Europe and North America; being openly racist became something to be ashamed of. Several countries (including the UK, which had ratified ICERD in 1969) brought in legal measures to deal with some forms of racism. The discourse in many fields of society shifted towards discussing the new racisms (see below), which are generally subsumed within the term 'institutional racism'—an expression coined by Stokely Carmichael and Charles Hamilton in their iconic book *Black Power: The Politics of Liberation* (1967, p. 4) and discussed more fully later in this chapter. Although institutional racism is the commonest of the new racisms, other terms carrying similar meanings tend to be used—such as structural racism, cultural racism, colour-blind racism and so on, all perhaps implying the notion that racism is not easily identified as being located in the actions or attitudes of a person, suggesting a concept of 'racism without racists' (Bonilla-Silva 2014, title). What has happened since about the early 1980s is that racist talk has been dropped so that any indication of racism, especially in conversation, is a matter to be ashamed of. Whether this was a fundamental change of popular ethos or mere 'political correctness'—an indication of politeness while real feelings are hidden—was never clear (see the section 'Rise of the political right' in Chap. 8 for further reference to political correctness).

Psychology and psychiatry were slow to adjust to a world in which adherence to principles of equality and human rights meant that active measures had to be taken to counteract racism—and to accept that racism would not just go away unless active anti-racist processes were put in place. Academic pursuits and clinical practice in the 'psy' disciplines did little more than pay lip service to being against racism and did very little to correct outdated ways of thinking about 'race' in their areas of study. This is typified by the massive study by Carleton Coon (1963), Professor of Anthropology at the University Museum in Philadelphia, which harked back to the sort of ideologies and divisions of humankind based on European race thinking (see section 'Race thinking' in Chap. 2)—a book described by Julian Huxley, knighted in 1958 and then President of the British Eugenic Society between 1959 and 1962, as '[a] valuable contribution to the evolutionary biology of man' (Coon 1963, cover). The legacy of race psychology of the first half of the twentieth century, especially the notions from the 'science' of eugenics (see the section 'Race psychology' in Chap. 3) continues to influence post-war psychiatry and clinical psychology.

4.2 American Social Studies

As sociology broke away from its attachment to anthropology in the early part of the twentieth century, the discipline, now increasingly called 'social sciences' (to indicate its claim of adherence to scientific objectivity), began to show some interest in social issues vis-à-vis minority groups in Western countries. A renowned study from the USA that focused on the effects of social conditions and the personalities of black people was the book *The Mark of Oppression* by Kardiner and Ovesey (1951). Abram Kardiner was an anthropologist and psychoanalyst and Lionel Ovesy was a psychoanalyst, and the title of their book indicated what they thought was the basic cause of many of the problems (including poverty and mental health issues) experienced by African Americans. It was essentially based on psychodynamic assessments (using psychoanalytic insights) of 25 case records of black people considered against a background of the history of African Americans in American society. The book discussed the cultural deprivation, lack of family cohesion and social disorganisation of the times of slavery and the harm that had resulted from racial discrimination in American society since emancipation; and it then went on to state that these experiences had resulted in 'depressed self-esteem' and 'self-hatred' within the black personality, partly dealt with by being 'projected' as aggression and anxiety. Kardiner and Ovesey go on to state blandly that the original (African) culture of African Americans having been 'smashed, be it by design or accident' (1951, p. 39), African Americans were virtually deprived of 'culture' and therefore living in a sort of cultural vacuum, their family life disorganised and the dominance of African-American women disturbing family cohesion. All this was presented as scientific study. Astoundingly, the authors claimed: 'There is not one personality trait of the Negro the source of which cannot be traced to his difficult living conditions. There are no exceptions to this rule … The final result is a wretched internal life' (1951, p. 81) (see Table 4.1).

The line of argument in *The Mark of Oppression* apparently resonated with what was commonly believed by American sociologists (who were mostly white people) and reflected the times in which it was written—the time of racial segregation in the USA, a few years before the civil rights movement. The deductions made about black people were clearly in line with what white Americans generally thought about African Americans—that black family life lacked cohesion and that black people were passive. The book acknowledged the racism that African Americans had been

Table 4.1 Theories of black racial inferiority

	Basic pathology	Effects on individual	Result
Old racism	Inborn characteristics	Underdevelopment	Small brains
	Biological	Inferior brain power	Low IQ
	Primitive culture		Passivity (needed domination)
New racisms	Tangle of pathology	Dysfunctional families	Deviant personalities
	Discrimination	Cultural 'vacuum'	Deviant behaviour
	Oppression	Wretched internal life	Poor perception of 'self'
	Primitive culture	Low self-esteem	Emotional underdevelopment
	Cultural deprivation	Self-hatred	Low IQ
		Cheery denial	Passivity (irresponsible)

subjected to for many years and attributed their social and psychological problems to their own pathological characteristics, thus stigmatising the oppressed. The message was that racism had resulted in black people having abnormal personalities—the implication being that *they* needed to change, to have treatment, for example.

Later studies in the USA of black families and culture were gathered together in a book by Daniel Patrick Moynihan (1965), *The Negro Family: The Case for National Action*, which informed American social policy and also influenced the thinking of psychologists and sociologists. In that book, later referred to as a 'report' called the Moynihan report, Moynihan described what he saw as a 'crisis' in urban (black) ghettos in terms of criminality, unemployment, educational failure and fatherlessness; and he focused very much on (what he saw as) dysfunctional or deviant black families. He explained that there was an unravelling of the black family, associated with growing rates of teenage pregnancy, dropping out of school and dependency on welfare—what he called the 'tangle of pathology' (Table 4.1). According to Moynihan, there were two sets of explanations for the crisis. One was the presence of certain cultural norms (of dependency, family organisation and crime) in the black community that were (according to Moynihan) the long-term legacy of slavery. The second was the high incidence of unemployment among black men, which reduced their desirability and practicability in the marriage market.

The message of the two books (*The Mark of Oppression* and the Moynihan report) was that African Americans had suffered long years of oppression because of their 'race' and that this had resulted in abnormal personalities and dysfunctional families. While earlier studies (see Chaps. 2 and 3) had argued that African Americans were inferior because of genetically determined small brains and low IQ together with their primitive cultures, these post-war studies, supposedly liberal in admitting to the prolonged racist oppression of African Americans, 'found' that they had abnormal personalities, dysfunctional families and other psychological difficulties and inadequacies because of oppression and discrimination. In both instances, African Americans were seen as inferior people, emotionally deficient, with deviant personalities and behaviours. The 'problem' was in *them* as individuals and families, not in *us* (white individuals and white society).

By acknowledging the extensive and prolonged racism in American society the Moynihan report, like *The Mark of Oppression,* appeared to be a liberal document, but both failed to recognise their own special type of racism, a subtle racism that concurred with white supremacy, and which was embedded in the way they were written and the implications contained in them. Both acknowledged racism in American society but both were based on a naive view of human development where negative experiences were assumed to lead to personality defects and pathologies in individuals and families. Judgements about family cohesion and the role of women in African-American societies were deductions made from a white perspective, which assumed that white families and white people were the norm. Also, another major failure was their lack of recognition that oppression might uplift as well as depress self-worth, and may promote as well as destroy communal cohesion, if sociopolitical conditions and power structures in the wider society allow that to happen.

A sociological approach that transfers the focus of emphasis from the oppression—racist oppression that was ongoing in this case—to the oppressed, inevitably has the effect of pathologising and stigmatising those oppressed, and letting the oppressors off the hook. Both reports, more explicit perhaps than the Moynihan report, appeared to place the onus on black families to make structural changes in family life, with an implication that white society would impose such changes if it did not come from within. Stokely Carmichael denounced the Moynihan report 'in a famous 1966 speech which popularized the slogan "Black Power" and signal[l]ed the radicalization of one segment of the civil rights movement' (Geary 2015, p. 119). The term 'institutional racism' was coined by Stokely

Carmichael to be taken up later (but defined rather differently) in the UK (Garner 2010) following public concern at the failure of London's Metropolitan Police to properly investigate a racist murder in 1993 (see 'The Macpherson report' in Chap. 6).

In a recent analysis of its legacy, Daniel Geary (2015), Professor of US History at Trinity College Dublin, has explored the legacy of the Moynihan report, analysing (with hindsight) the controversies surrounding it. A significant point made by many critics was that, by stressing 'the "virus" of individual racial prejudice and the effects of historical oppression on African American social structure', Moynihan underplayed the importance of 'ongoing racism embedded in American institutions' (Geary 2015, p. 80). This allowed neoconservative forces and segregationists to oppose positive action to counter racism. Geary states that William Ryan, an activist from the North 'famously accused Moynihan of "blaming the victim" by focusing attention on African American families rather than the system that oppressed them', thereby fuelling 'a new racism' (p. 80). Ryan's criticism was supported by civil rights organisations and the Moynihan report became an embarrassment to the Democratic government of Lyndon Johnson; but once Richard Nixon became President of the USA, Moynihan joined the White House staff and is alleged to have advised Nixon in 1970 that 'the issue of race could benefit from a period of "benign neglect"' (Geary 2015, p. 201).

Unfortunately, American ideas about black families (as described in *The Mark of Oppression* and the Moynihan report) were taken as fact when they crossed the Atlantic, becoming evident in British research in the form of negative images that developed of African-Caribbeanand Asian British families. It should be noted that in the UK, the term 'Asian' refers to people of South-Asian origin, with those of Chinese and Japanese origin (although relatively few in number) being referred to as 'British-Chinese' and 'British-Japanese'; while in North America 'Asian' means (mostly) people of Chinese, Japanese *and* South Asian origin as well as people whose ancestry could be traced to the South Pacific Rim of nations. The alleged inferiority of the black person's brain, personality and intelligence, cited in former times (when 'black' included both black and brown-skinned people, see Chaps. 2 and 3) was supplemented in the 1960s and 1970s by 'culturalist' theories, which referred to the allegedly defective family and kinship systems, marital arrangements and child-rearing practices of black communities—'a picture of "pathological" black households … which fits in only too well with common-sense racist imagery' (Lawrence 1982, p. 95). According to Errol

Lawrence (1982), African-Caribbean people in Britain were seen as having a family life that was weak and unstable, with a lack of a sense of paternal responsibility towards children; and Asian families were seen as strong, 'but the very strength of [British-] Asian culture… [was seen as]… a source of both actual and potential weaknesses' (1982, p. 118). The American Moynihan report called the black African-American family 'a tangle of pathology'; in the UK, a Select Committee on Race Relations (1977) reported a connection between the problems of African-Caribbean British families and family life in the Caribbean, the latter being seen as unsuited to British society. In the field of psychiatry, the notion of pathological family structures ('tangle of pathology') causing schizophrenia among African-Caribbeans—this time as a reason for the high rates of schizophrenia being diagnosed among British black people—surfaced in the British press (Lewin 2009) as a result of comments made by researchers at the Institute of Psychiatry (IOP) that recommended social engineering of black family structure (see 'Racist conclusions of psychiatric research' in Chap. 8).

4.3 Black Protest in the UK

In 1981, race riots occurred in Britain (see Solomos et al. 1982; Olusoga 2016), the first serious civil disturbances involving 'race' since the post-WWII immigration. 'These riots were called "uprisings" [because] [t]hey were fought by young black people in response to years of systematic persecution and prejudice' (Olusoga 2016, p. 517): the latter included excessive use, against black youth, of the police's legal power to stop and search people in the street. The riots affected Brixton in London, Handsworth in Birmingham, Chapeltown in Leeds and Toxteth in Liverpool, resembling *in toto* the 1960s race riots in the USA during the civil rights movement. Darcus Howe (2011), a leading activist for the rights of black people, has described the lead-up to the Brixton riots: in January 1981 there had been a fire during a sixteenth-birthday party in a house in New Cross, South London, attended by mainly African-Caribbean youngsters. The fire, which left nine dead—and four more died two weeks later—was widely suspected to have been the work of an arsonist, but the police took little interest in finding its cause, instead subjecting the young people who had attended the party to aggressive questioning. In the decade before this event, London's black youngsters has been at the forefront of demonstrations and campaigns against so-called 'sus laws' (that empowered police to stop and search people on suspicion) being used disproportionately to stop and search black youngsters; and white politicians had repeatedly blamed black people for causing trouble on the streets. After the fire, the local black community gathered together—the end result being the famous march of 20,000 black people from Deptford to Hyde Park. Two weeks later, the police launched Operation Swamp, enabling them to indiscriminately stop and search (under sus laws) black youth in Brixton. The riots started with the slogan 'This is for New Cross'. Riots in other British cities followed in the wake of the Brixton uprising.

There had been race-based civil disturbances in the UK before 1981 but they consisted of white gangs attacking black people in Glasgow, Liverpool and London (in 1919) and parts of London in 1948 and 1958 (see Olusoga 2016). The Tory government which was in power in the UK in the 1980s appointed Lord Scarman to investigate the causes of the Brixton riots, and there were less extensive investigations into the riots in other cities (for example, the inquiry into the Handsworth riots by Silverman 1986). Scarman's report (Home Office 1981) into the Brixton riots denied the existence of racism among police officers except for a few 'bad apples' and blamed the riots on a mixture of the attitudes of young black people, the racial disadvantages they suffered in society, and the inexperience of police officers. The remedies recommended by Scarman consisted of positive action to meet (what were called) 'the special problems and needs of ethnic minorities' (1981, p. 108)—meaning that their culture was different from that of white people. As a result, there were active efforts to promote the religious and cultural identities of ethnic minority groups but some authorities also encouraged psycho-educational approaches, such as racial awareness training (RAT), for white people—RAT being an approach to addressing personal prejudice that resulted in overt discriminatory practice.

Looked at with hindsight, this report set back progress in tackling racism (Bourne 2001) by presenting disadvantage as 'some sort of handicap' (2001, p. 11) suffered by black people. The Scarman report 'effectively … shifted the object of anti-racist struggle from the state to the individual, from changing society to changing people, from improving the lot of whole black communities, mired in poverty and racism to improving the lot of black individuals' (2001, p. 12). RAT training was also ineffective. People attending RAT courses often shrugged them off as not applicable to them since they were not racist; and those 'trained' often felt absolved from having to do anything further about racism. I remember how social workers sent to attend racial awareness courses returned to practice convinced that they were no longer racist, as if they had been immunised against contracting an infectious disease. So racism carried on until a wake-up call arrived in 1999 in the shape of a damning report on why the police in London had failed to properly investigate the murder of a black teenager—see the 'Macpherson report' section in Chap. 6.

4.4 Definitions of Racism and Race

The lack of precision in defining racism has resulted in much argument and confusion about its nature in the world of today—to the extent of questioning its very existence as a problem, just as 'the absence of a clear "common sense" understanding of what racism means has become a significant obstacle to efforts aimed at challenging it' (Omi and Winant 1994, p. 70). Omi and Winant (1994) argue that a part of the problem in understanding the nature and reality of racism today arises from its being seen as either an ideological phenomenon (with its own beliefs, attitudes, doctrines and discourse) that results in unequal and unjust practices, or as a structural phenomenon (with the attendant economic stratification, residential segregation, and institutional forms of inequality) that then gives rise to ideologies of privilege—white privilege (see 'Privilege and power' in Chap. 7). They explain that 'ideological beliefs have structural consequences and social structures give rise to beliefs', so that racial ideology and (racist) social structure 'mutually shape the nature of racism in a complex, dialectical and overdetermined manner' (1994, pp. 74–75). In fact, an inextricable linkage existed even in the development of the overtly racist plantation slavery of the USA between racism on the one hand, and on the other, factors such as the shortage of cheap labour and the elaboration of juridical and property rights (Hall 1980). Further, legacies of racism have 'entrenched massive inequalities between groups of people designated as races … [a]nd these differences continue to be perpetuated by ongoing practices' (Durrheim et al. 2009, p. 198).

Omi and Winant (2015) issued an updated second edition of their classic book *Racial Formations in the United States*, in which they clarify the idea of 'race' (closely linked of course to the understanding of racism and institutional racism): 'To say that race is socially constructed is to argue that it varies according to time and place' [and that] 'race is *ocular* in an irreducible way … [in that] human bodies are visually read, understood, and narrated by means of symbolic meanings and associations' (p. 13, emphasis in original). Thus they reach the following definition of race—a *social* definition: '*Race is a concept that signifies and symbolizes social conflict and interests by referring to different types of human bodies*' (Omi and Winant 2015, p. 110, emphasis in original). The final part of their definition (the focus on human bodies) is important because that takes on the history of how the notion of race came about (see 'Exploration, colonialism, race-slavery' in Chap. 1) and also takes into account how racism has been experienced ever since.

In his book *Racist Culture,* David Goldberg (1993) argues against considering racism as a homogenous phenomenon: 'It follows that there may be different racisms in the same place at different times; or different racisms in various different places at the same time; or, again, different racist expressions—different that is, in the conditions of their expression, their forms of expression, the objects of their expression, and their effects—among different people at the same space-time conjuncture' (1993, p. 91). Thus racism during the era of slavery in America differs from post-slavery segregationism, and each of these from current expressions of racism in the USA. Racism in South Africa during the times of apartheid differs from that expressed there through inherent economic inequalities in the post-apartheid era. Nineteenth-century British racism in the colonies differs from current manifestations of racism in the UK. The most reasonable and realistic approach is to regard racism as something that does not *just* exist as an abstract concept; its importance lies in its relevance to social relations between people. In other words, racism cannot be adequately explained in abstraction from social relations, nor can it be explained by reducing it to them (Hall 1980).

Essed (1990) proposed the concept of 'everyday racism', which is very much about the personal experience of racism in the course of day-to-day interactions between people. The people who exhibit racism are not necessarily overtly (racially) prejudiced, although if one examined their attitudes in some depth racist attitudes may be uncovered. Their apparently unwitting racism may be expressed through ways of behaving and socialising. But if this approach is taken further, racism may be manifested in social and political systems because people unwittingly collude in supporting it, possibly because they gain from doing so (see also the definition of institutional racism as it developed in the UK, which is described in the 'Macpherson report' section in Chap. 6). Wellman (1977) argues in *Portraits of White Racism* that once racial prejudice is embedded within the structures of society, individual prejudice is no longer the problem —'prejudiced people are not the only racists' (1977, p. 1).

Carmichael and Hamilton (1967) began their book *Black Power: The Politics of Liberation* by referring to two 'closely related forms' of racism in the USA—'individual whites acting against individual blacks, and acts by the total white community against the black community' (p. 4):

> When white terrorists bomb a black church and kill five black children, that is an act of individual racism, widely deplored by most segments of the society. But when in that same city—Birmingham, Alabama—five hundred black babies die each year because of the lack of proper food, shelter and medical facilities, and thousands more are destroyed and maimed physically, emotionally and intellectually because of conditions of poverty and discrimination in the black community, that is a function of institutional racism. When a black family moves into a home in a white neighborhood and is stoned, burned or routed out, they are victims of an overt act of individual racism which many people will condemn—at least in words. But it is institutional racism that keeps black people locked in dilapidated slum tenements, subject to the daily prey of exploitative slumlords, merchants, loan sharks and discriminatory real estate agents. The society either pretends it does not know of this latter situation, or is in fact incapable of doing anything meaningful about it. (Carmichael and Hamilton 1967, p. 4)

The term 'institutional racism' became popular in the UK much later after it was cited in an official report (Home Department 1999) as the reason for failures by London's Metropolitan Police to properly investigate a racist murder in 1993 (see 'The Macpherson report' in Chap. 6). For the purposes of the report, institutional racism was defined in such a way as to connect with the sort of racism that could be legally defined (see Garner 2010, pp. 102–116 for discussion of confusion caused by use of the same term in different senses). It is in effect the most frequently referenced new form of racism that has emerged and so the expression is sometimes used in this book as a sort of generic term to cover all types of subtle racism, in the American sense as well as the British. Today racial prejudice at a personal level exists alongside, and interacting with, these other forms of racism. Further discussions of racism and the notion of 'race' are covered in 'Racialisation' in Chap. 5.

The term 'ethnicity' has become popular in the UK as an alternative to 'race', partly because the latter carries baggage from the past and partly because 'ethnicity' has a broader scope in capturing what is meant by personal identity. In the sense in which it is used in the UK—it may be used differently elsewhere—it means both race and culture. In practice, for example in ethnic monitoring of populations, information is obtained by asking someone about their self-given identity, their sense of belonging to a group, either cultural or racial or both. The term is used mainly for (ethnic) monitoring of populations in order to detect inequalities or differences between groups—in which case it is often assumed to be the same as 'race'. Sociologists tend to see ethnicity as a comprehensive indication of how a person sees themselves as a unique individual. Stuart Hall (1992) states: 'The term ethnicity acknowledges the place of history, language and culture in the construction of subjectivity and identity, as well as the fact that all discourse is placed, positioned, situated, and all knowledge is contextual' (p. 257). Ethnicity is more liable to change over time; sociologists refer to 'new ethnicities' as having emerged in British society during the 1980s and 1990s (Cohen 1999) as a result of African, Caribbean and Asian cultures being perceived as 'different' to the majority 'white' culture and, more importantly, people seen as belonging to these cultures being perceived as 'different' racially. In a multiracial and multicultural society categorisation in ethnic terms emerges in complex ways through various pressures, social, political, economic and psychological. Perceived or strongly felt cultural ties and/or perceptions of being part of a 'race' may exist—and the strength of any of these may be influenced by a variety of pressures. Among the social forces is the pressure arising from racism that drives together those people perceived (by others) as being racially similar to each other. Ethnicity that is promoted and crystallised by social forces has been called 'emergent ethnicity' (Yancey et al. 1976). This is the ethnicity that is of practical importance in most societies where there is racial and cultural diversity.

4.5 New Racisms in the UK

In his analysis of the connection between racism and colonialism (from a 1965 speech published two years later), Frantz Fanon (1967) talked of 'vulgar racism in its biological form' (p. 35), corresponding to a period of crude exploitation, which then changes into 'cultural racism'—'a more sophisticated form [of racism] in which the object is no longer the physiology of the individual but the cultural style of a people' (McCulloch 1983, p. 120)—hence the term 'cultural racism'. The idea that racism is embedded deeply in the culture of a society leads to the notion that it is incorporated in that society's structures (hence 'structural racism') and/or its social and political institutions; this situation is often referred to generally as institutional racism.

In the late 1950s, I found that it was perfectly in order for doctors at the hospital that I was then working in, and where about 10% of the junior doctors were 'coloured', to refer to an informal bar run by junior doctors and which was based in a room in the hospital as the 'colour bar'—a no-go area for 'coloured doctors' (the term 'coloured' was then considered more polite a term than 'black' to describe black or brown-skinned people). In fact the 'colour bar' was spoken of fairly freely as an established fact of life. I recall how difficult it was for 'coloured' students and professionals to get accommodation in 'respectable' areas in London and Cambridge. In seemingly 'respectable' towns, such as the town of Epsom where I worked as a trainee psychiatrist in the early 1960s, I was often deprived of service at restaurants by being asked to wait for a table indefinitely (until I left); and in less 'reputable' areas I was often told abruptly that there was no room when clearly there were empty places—see Chap. 1 for further similar examples. Legislation outlawing race discrimination came on the scene in 1965 (Hepple 1966) and was strengthened in the Race Relations Act (1968). By the mid-to-late 1970s, I rarely encountered the type of overt prejudice in public places that had been common in the 1960s (but for example of other types of racism see 'Employment in the mental health system' and 'Institutional racism in the Department of Health (DOH)' in Chap. 7).

From about the early 1980s onwards, it was clearly impolite, if not actually insulting, to express racist ideas in writing or in general discourse. It is likely that this reflected a desire on the part of many people to avoid giving offence to others, but it also meant that racism was (as it were) driven underground—it was *felt* (and perhaps believed in) but not expressed because it was incorrect—or immoral—to do so. It was in this context that the notion developed that being 'politically correct' in speech or written word meant that the person concerned wants to signal moral superiority (Weigel 2016). The racism that persisted, sometimes called 'a new racism based on arguments about cultural difference' (Cohen 1992, p. 67), largely took over the public discourse, including within it concepts such as 'institutional racism' and similar notions (see 'Definitions of Racism and Race' earlier this chapter). As racism became more complicated and varied than it used to be, it was less obvious but perhaps more dangerous for this very reason. And yet, the former overt racism ('old-fashioned' and unashamed—see Chaps. 2 and 3) in the form of direct personal prejudice still continued, seemingly existing just below the surface to become explicit and obvious whenever circumstances enabled or provoked its appearance (see 'Rise of the political right' in Chap. 8 and 'New era of unashamed racism?' in Chap. 9).

4.6 Racist IQ Movement

In the world of psychiatry and clinical psychology, 'race psychology' (Chap. 3) rumbled on through the twentieth century in spite of political and social upheaval in Europe, partly caused by the rise of aggressively overt racism, which led to the Nazi movement and WWII. The next few paragraphs consider significant locations where racism in the 'psy' disciplines had important influence post-WWII—the fields of intelligence studies, of the nature of 'mind' and mental illness, and of the social and family dynamics that affect mental health.

The overt racism expressed in the scientific racism of the eighteenth and nineteenth centuries (Chaps. 2 and 3) was seldom seen in scientific publications after WWII, but it was soon obvious that racism was far from dead in the 'psy' disciplines, although these fields claimed to depend on scientific (and hence objective) evidence for their research, policies and practices—and science was seen as unbiased. In 1969, a significant paper by Arthur Jensen (1969), Professor of Educational Psychology at the University of California, appeared in the *Harvard Educational Review* and was well received by many psychologists. Jensen proposed that differences between black and white people (in the USA) in scores on IQ tests were genetically determined. Further, Jensen postulated two categories of mental ability—abstract reasoning ability, characteristic of white people, and rote learning, characteristic of black people. British psychologist and Professor of Psychology at the prestigious IOP in London, Eysenck (1971, 1973), supported Jensen's views, and the basic premises of the theory were widely accepted by bodies responsible for education on both sides of the Atlantic, although they were criticised by some psychologists like Watson (1973), Kamin (1974) and Stott (1983). It is likely that the arguments of Jensen and Eysenck may have helped the partial revival of scientific racism against the trends in scientific work at the time.

In an important booklet *Psychology, Racism and Fascism*, Billig (1979), a professor of social sciences, analyses the connections between the rise of fascism in the 1970s and the psychological theories of race that grew up at the time. He recalls that in the 1930s, '[seemingly] "respectable" scientists' (principally geneticists, biologists, physical anthropologists and psychologists) contributed to the growth of *Rassenkunde* (literally "Race- science")' which supported the rise of Nazism in Germany (p. 5). The message of race science was continued after WWII in *The Mankind Quarterly*, 'a journal with an impressive scholarly appearance' and H. J. Eysenck on its Advisory

Board, which was established in 1960 in Scotland and edited by Professor R. Gayre, a physical anthropologist trained in Edinburgh. From its earliest issues, this journal 'provided a platform for former colleagues and heirs of Nazi racial theorist Hans Gunther' (pp. 11–15). Billig quotes the evidence given by Gayre (as an expert witness) in defence of members of the Racial Preservation Society, who were charged in the UK with an offence under the (British) Race Relations Act (1968); he 'maintained that scientific evidence showed that blacks "prefer their leisure to the dynamism which the white and yellow races show"' (p. 12). Among several eminent scientists involved in *The Mankind Quarterly* during the 1970s were Robert Kuttner, an American biochemist, Professor Henry Garrett, a past President of the American Psychological Association and Chair of Psychology at Columbia University and Professor H. J. Eysenck of the (British) IOP. The racist IQ movement continued in books such as *The Bell Curve* (Herrnstein and Murray (1994) and carries on into the twenty-first century, supported by academics such as J. Phillipe Rushton, quoted by Richards (2012).

The terrible consequence of racist theories put forward by important and powerful people who claim to be scientists is that these theories can be self-fulfilling. For example, as Geary (2015) argues, once the proposition 'that hereditary differences in intelligence explained racial disparities' (p. 197) (highlighted in the Moynihan report—see the section 'American social studies' earlier in this chapter) provided a convenient theoretical bolster for holding back on liberal social policies to improve educational opportunities for African-American children, that very action/inaction (in the 1970s and 1980s) of failing to improve opportunity led to African-American children continuing to score lower in IQ tests when compared to white children. When (for example) a psychology report issued by a division of the British Psychological Society (BPS) contains language that may reflect institutional racism (see 'Racism of a psychology report' in Chap. 8), the discipline of clinical psychology may well be seen as racist, and psychological therapies avoided. And when black people are persistently over-represented as 'schizophrenic' (see 'Ethnic issues in mental health services' in Chap. 5) the system of psychiatry that allows this to happen is justifiably seen as institutionally racist and people who may require help may be put off attending mental health services.

4.7 Alleged Mentality of Black People

Three distinct views about the minds of (culturally) non-Western peoples, usually identified in racial terms, were discernible during the development of psychiatry. In the mid-eighteenth century, Rousseau's concept of the 'Noble Savage' proposed the view that 'savages' who lacked the civilising influence of Western culture were free of mental disorder; later, in the late-eighteenth and nineteenth centuries, Daniel Tuke (1858) and Maudsley (1867, 1879) in England, Esquirol (cited by Jarvis 1852) in France and Rush (cited by Rosen 1968) in the USA voiced similar views, expressed most firmly by J. C. Prichard (1835) in his *Treatise on Insanity*: 'In savage countries, I mean among such tribes as the negroes of Africa and the native Americans, insanity is stated by all ... to be extremely rare' (p. 349). But Aubrey Lewis (1965), well-known British psychiatrist, writing in an introduction to one of the earliest books on transcultural psychiatry, pointed out that a second, somewhat different, stance was also evident in Europe at about that time; namely, the view that non-Europeans were mentally and morally degenerate because they lacked Western culture. A third viewpoint was voiced in the USA by psychiatrists arguing for the retention of slavery: epidemiological data based on the Sixth US Census of 1840 (Anon 1851) were used to justify a claim that the black person was relatively free of madness in a state of slavery, 'but becomes prey to mental disturbance when he is set free' (Thomas and Sillen 1972, p. 16). The underlying supposition was that the inherent mental inferiority of Africans justified their enslavement.

Although the 'Noble Savage' viewpoint idealised non-European culture in some ways and the 'degenerate primitive' attitude vilified it, both approaches sprang from the same source—a racist perception of culture which supposed that European culture alone, associated with white races, was civilised; and the culture of black people, being primitive, rendered them either free of mental disorder or inherently degenerate. These views, determined by a warped perception of the lives of black Americans—or more correctly, determined by a need to justify slavery—left no room for cultural considerations at all; in fact an assumption that black Americans *lacked* a culture was implicit in the way these ideas developed. The underlying theoretical question that was being addressed in the discussions about 'civilisation' and mental disorder noted above was akin to more recent discussions about the universality of schizophrenia—reviewed by

Richard Warner (1985) and Fuller Torrey (1987). As then, the matter is currently confused by racism; this more recent debate resonates with the fear in the USA during slavery era that black lunatics might become more numerous as years go by (see the section 'Mental pathology and the construction of race-linked illnesses' in Chap. 3).

One of the earliest observations reported by a psychiatrist about mental illness among Asian people was the claim by the German psychiatrist Kraepelin (1913) that the people of Java, now a part of Indonesia, seldom became depressed, and that when they were, they rarely felt sinful. Kraepelin (1920) perceived these differences in terms of genetic and physical influences rather than cultural ones—a reflection not only of the biological orientation in German psychiatry at the time, but also of the acceptance of racial explanations for cultural difference. In fact, Kraepelin (1921) saw the Javanese as 'a psychically underdeveloped population ... [akin to] ... *immature* European youth' (p. 171, emphasis in original), and for explanations in racial difference. Sashidharan (1986) believes that Kraepelin's notion 'became detached from mainstream psychiatry over the next few decades and gradually orientated itself around emerging ideas from anthropology and psychoanalysis' (p. 168). Since both these disciplines also carried racist ideas about culture, the 'transcultural psychiatry' that arose contained within it a racist tradition which later adherents tried to extrude. Yet, theories have emerged in modern psychiatry about culturally determined differences (in brain function, emotional differentiation, personality defects, family life, and so on), all of which theories often harbour racist doctrines (Fernando 1988).

The WHO, established soon after the end of WWII, sought to address issues in mental health worldwide, especially in countries that had been underdeveloped during colonial times. J. C. Carothers, a British colonial psychiatrist who had been the superintendent of a mental hospital in Kenya, then a British colony in East Africa, was invited to produce a report on 'The African Mind'. The monograph that Carothers produced, *The African Mind in Health and Disease* (Carothers 1953), was hailed as the standard text for understanding Africans and remained so for several years. Earlier, Carothers (1947) had proposed several astounding explanations for the 'peculiarities of primitive African thinking', which he saw as 'inherited' (p. 581). He deduced that 'the rarity of insanity in primitive life is due to the absence of problems in the social, sexual and economic spheres', while contending that the 'African may be less heavily loaded with deleterious genes than the European... [because] ... natural selection might be expected to eliminate the genes concerned more rapidly in a primitive community' (pp. 586–587). He commented on what he saw as a lack of pressure on Africans because they (allegedly) had no long term aims in life and he noted the apparent lack of depression among them: 'Perhaps the most striking difference between the European and African cultures is that the former demands self-reliance, personal responsibility, and initiative, whereas there is no place in the latter for such an attitude' (p. 592). Four years later, Carothers (1951) took his studies much further. He made deductions about the 'neurophysiological basis of African thinking' to conclude a 'striking resemblance between African thinking and that of leucotomized Europeans' (p. 12).

> The African attitude implies that, apart from certain swift and almost automatic responses and inhibitions, he can do what he likes from moment to moment and feels little need to think of the future or indeed of any other than the immediately presenting aspect of the situation. So he feels free to exercise his most egotistic and emotional impulses (within well-defined limits) and such mental organisation as he evinces is imposed from without and not self-developed. He is hardly in fact an individual in our sense of the word, but a series of reactions. (pp. 33–4)

In his monograph for the WHO, Carothers (1953) reiterated the racist views propounded earlier (see 'Mental pathology and the construction of race-linked illnesses' in Chap. 3) that the brains of African and American black people were inferior to those of Europeans, conveniently forgetting that those findings had long been discarded. Carothers' claims about Africans were, in effect, a compendium of racist stereotypes of black people, referring to their (alleged) failure in psychological development after puberty, 'lack of spontaneity, foresight, tenacity, judgement and humility' (p. 87) and so on. Although he referred frequently to 'culture' as the basis of all their (alleged) peculiarities, the discussion and presentation in his treatise, with references to black people in both Africa and America as similar, if not the same, both culturally and racially, clearly indicate the racist and not cultural nature of his assumptions. The fact that Carothers' monograph for the WHO was widely quoted as an authoritative treatise on the psychology of Africans while at the same time the UNESCO document on race (see 'Transformations after WWII' earlier in this chapter) was officially hailed as a standard for a post-war era, is indicative of the power of racism in Western culture and the degree to which racism is embedded in the 'psy' disciplines. Although Carothers' work is seldom quoted now, the ideologies he promoted remain in that they have fed into theories and clinical work with black and minority ethnic (BME) people in the UK.

Like Carothers in the British colony of Kenya, the most significant psychiatrist in the French colonial possessions in North Africa (the Maghreb) was Antoine Porot. In 1918, Porot 'framed the "North African" [natives of the region] as inherently puerile and incapable of coping with the realities of modern civilization'; and in 1925 he produced spurious data showing the African mind as being 'primitive, criminally impulsive, and intellectually feeble' (Keller 2007b, pp. 18–25). Fanon (1967) quotes Porot to have stated in 1935 at the Congress of Mental Specialists and Neurologists that 'the native of North Africa, whose superior and cortical activities are only slightly developed, is a primitive creature whose life, essentially vegetative and instinctive, is above all regulated by his diencephalon' (p. 243)—the diencephalon is a developmentally primitive part of the brain. According to Keller (2007a), psychiatry in the Maghreb under the leadership of Porot eagerly took up somatic 'treatments' that were considered experimental at best in France. In the 1940s, ECT (electroconvulsive therapy or 'electroshock') became the 'bedrock of therapy in the Maghreb' and Professor Porot performed over 200 lobotomies between 1947 and 1954 (Keller 2007b, pp. 29–30). Clearly, not only was Porot misguided and his work institutionally racist—and it is likely that Porot himself was personally racist —but he caused a lot of suffering and social damage.

Porot's work, together with 'observations' [*sic*] by other French psychiatrists working in North Africa on the 'primitive mentalities' and 'criminal impulsiveness' of North Africans, became incorporated into the 'discipline of ethno-psychiatry [that] informed education and professional discrimination against Muslims, shaped discourse about immigration into France and provided the essential background for the French army's psychological warfare programs during the Algerian struggle for independence' (Keller 2007a, p. 7). Clearly, French 'ethnopsychiatry' informed the practice of psychiatry in the Maghreb until its liberation between 1956 and 1962 and the founding of the nation states of Tunisia, Morocco and Algeria. It is likely that the racist notions promoted in French 'ethno-psychiatry' influence psychiatric practice in France today vis-à-vis French citizens of North African origin. However, this is not a topic that has been researched or discussed at all in the contemporary literature, unlike the case of the UK, where a large literature exists on racism in British psychiatric practice.

4.8 Racism in Cultural Research

Cultural differences in the vulnerability to depressive illness and the way depressive feelings may be expressed in different cultural settings has been (and is) a popular theme in psychiatric research. In a review of culture and depression, Bebbington (1978) uses the term 'primitive cultures' as meaning non-Western cultures and, more significantly, argues for 'a provisional syndromal definition of depression as used by a consensus of Western psychiatrists against which cross-cultural anomalies can be tested' (p. 303). In other words, the 'depression' of non-Western peoples is hailed as an 'anomaly' and the paper indicates that these so-called anomalies are found among black Americans, Africans, Asians and 'American Indians'—again, racial categories and a racist conclusion. Such observations are reflected in the apparent rarity of depressive illness being diagnosed in Africa until well into the 1950s. Carothers (1953) (referred to earlier) was one of the foremost among many (white) psychiatrists who claimed that depression was rare among (black) Africans. He was in tune with the general tenor of the time in attributing his alleged observation to the absence of 'a sense of responsibility' (among black people) (p. 148). Raymond Prince (1968) has noted that, since 1957 (the year of Ghana's independence) journal papers had appeared reporting that depression was not rare but common among Africans. He observed that 'the climate of opinion' determined the change: 'In the Colonial era, depressions should not be seen and named because Africans were not [perceived as] responsible' (p. 186). In other words, Carothers found depression to be rare in Africans *because* he saw Africans as lacking a sense of responsibility, rather than vice versa. There were other racist ideas in psychiatry about African mentality too—and many still influence psychiatric practice today. I recall being taught (when training in psychiatry in the UK) at a seminar at the IOP (then considered a centre of excellence) that mental illness in people from Asia and the West Indies often presented as an undifferentiated 'primitive psychosis' in that the symptoms were difficult to delineate clearly because of *their* particular mentality—and that was in the 1960s in London.

One theory propagated in the 1970s and now quoted in psychiatric texts is concerned with variations in the 'differentiation of emotions'. This study from the prestigious British IOP reported that people from 'developed countries showed a greater differentiation of the emotions than [did people from] developing countries', with black Americans resembling the latter in this respect (Leff 1973, p. 304). The collection of data was somewhat peculiar. The 'data' on ability to express emotions had been deduced from data collected for the International Pilot Study of Schizophrenia (IPSS) (WHO 1973), on the extent to which people had been judged as *experiencing* emotions; and then Leff had added supplementary data from a US-UK study on black Americans and white Americans (Cooper et al. 1972). But more seriously, the conclusion arrived at had racial undertones, especially when Leff (1977) claimed that his theory represented an 'evolutionary process'. There was a (racist) slippage of meaning whereby people from *industrially* underdeveloped countries were seen as *culturally* underdeveloped; and black African Americans in the USA were equated with 'underdeveloped' people from industrially underdeveloped countries.

A study by a team from the Social Psychiatry Unit at the IOP (Bebbington et al. 1981), comparing 'psychiatric disorder' among people of (black) West Indian origin with those of (white) Irish origin and indigenous British-born people (white non-Irish people), found a lower incidence of depression and anxiety among black African-Caribbean people, when compared to the other two groups. In the discussion as to why this was so, the researchers reported a 'clinical impression' that (black) West Indians 'respond to adversity with cheery denial' and attributed their 'relative immunity to minor disorder in terms of this cultural characteristic'—the Irish citizens 'seemed much more readily aware when things were going badly' (p. 51). In other words, the researchers explained *lower* incidence of depression and anxiety among black people by postulating a (pathological) 'cheery denial' (see Table 4.1), rather than, say, a positive skill such as 'stiff upper lip' (a quality that would undoubtedly have been attributed to a group seen as white). It is important to state at this stage that it is not necessarily individual (racial) prejudice—the prejudices of individual research workers at IOP or anywhere else—that one can blame for the racist conclusions of research. Racism was inherent in the way they carried out their work—the traditional colour-blind approach of scientific (objective) research; the failure to include ways of excluding racial bias into research methodology; lack of sensitivity to pervasive racism and racist ideologies in the society in which the research is being carried out; the biases of researchers in drawing conclusions, and so on. All this adds up to an 'institutional racism' that is embedded in psychology and psychiatry.

4.9 Conclusions

Looking back on the period before WWII, it appears that race thinking and racist notions were uppermost when differences were seen to exist between groups of people (or even individuals) who hailed from different cultural backgrounds or different nationalities—racist explanations were the norm, racist ideas were not frowned upon and were perfectly respectable. But *after* about the 1950s, as racism became unfashionable and then something to be ashamed of, racist notions were seen as serious misconceptions' that would eventually disappear as people became enlightened, probably as a result of liberal education. And, with a twist of retrospective memory, racism was even claimed as being contrary to (an imagined) set of 'European values' held since ancient times. In other words, the status of racism changed; it lost its pride of place in European thinking soon after WWII, but until then had been *fundamental* to European culture.

Action to suppress blatant racism was taken politically in many European countries during the 1970s and 1980s. However, subtle and indirect forms continued, and became active as overt racism from time to time. Also to some extent, the persistence of racism was obscured by political correctness in the use of language, meaning 'language … [that reflect[s] an increasingly diverse society—in which citizens attempt to avoid giving needless offence to one another' (Weigel 2016). Since institutional racism is the easiest form of the new racisms to recognise, the term 'institutional racism' is often used to cover many forms of new racisms.

In the case of the 'psy' disciplines in the UK, overtly racist notions were still very prominent in the UK well into the 1980s. However, there was a shift after this time, resulting (it seemed then) from black people themselves striving for equality by means of political action—for example in challenging police brutality and psychiatric racism—and supplemented by writings of black people on the effects of racism in general, for example classics by Homi Bhabha (1994), Paul Gilroy (1987) and Stuart Hall et al. (1978), and critical literature in the field of psychiatry and (to a much lesser extent) clinical psychology. However, widespread racism still prevails which, together with (and interacting with) cultural insensitivity in the disciplines, causes suffering and inequalities in mental health service provision in the NHS, a system that is otherwise held in high regard by most British people.

In the late 1980s, many BME people (and I was one) thought that the UK would progress over the next ten or fifteen years to reduce the extent of racism in the 'psy' disciplines to one that we could live with—few BME people thought that racism could ever be abolished. How wrong we were! One could visualise racism today as an influence that permeates through systems, ideologies and many different facets of social functioning; it affects ideas and attitudes informing discourse, both public and private, and professional practice in many different fields including clinical psychology and psychiatry. So in British society, racism—or perhaps racisms—takes/take diverse forms. The 'everyday racism' described by Essed (1990) (see earlier in 'Definitions of racism and race') merges into racial harassment and racial attacks, which are still rife in some parts of the UK and the continent of Europe. Racism in the mental health system continues, partly because the professional bodies involved have failed to take anti-racist action in their disciplines or they do not recognise the problem's existence. Although it is true to say that overt racial discrimination at a personal level—'street racism'—declined in the UK during the 1980s and 1990s, it continues to be experienced in varying degrees by black and brown-skinned people in their everyday interactions with others and it is implemented in subtle ways through British institutional processes—'institutional racism'. All forms of racism have been increasing since the early 2000s, some more than others, and are experienced in different ways in different locations.

The main lessons for the mental health field that come through are about the positive results of the struggles of black people during the many years of slavery, colonialism and discrimination in (predominantly white) Western societies; about the richness and variety of black and Asian cultures that have developed in the UK and USA; about the interaction and melding together of cultures; about the changing nature of racism—how it changes its form but is also opposed and being gradually ground down, albeit with ups and downs; about the forging of new identities and ethnicities; and about the struggles against racism in the 'psy' disciplines. Unfortunately mainstream psychiatry and psychology have so far failed on the whole to fully take on board the insights offered by the progressive thinking that flooded the British and American scene at the end of the twentieth century. This is the fault of the continuing racism embedded in clinical psychology and psychiatry, which exists in spite of an extensive critical literature in the 'psy' disciplines that both addresses old-fashioned processes of racial oppression and discrimination and the newer forms of racism, especially the ways in which racialisation has complicated the issues involved. The next four chapters explore further aspects of racism prevalent today, and how racism manifests itself in a changing world that appears to be getting more dangerous—in the 'age of anger' (Mishra 2017)—and the book concludes in Chap. 9 by considering the future we face, speculating on what may happen in the field of mental health and in the two 'psy' disciplines in the years to come.

REFERENCES

Anon. (1851). Startling Facts from the Census. *American Journal of Insanity, 8*(2), 153–155.

Arendt, H. (1958). *The origins of totalitarianism*. Cleveland: World Publishing Company.

Arendt, H. (1970). *On violence*. Orlando: Harcourt Books.

Bebbington, P. E. (1978). The epidemiology of depressive disorder. *Culture, Medicine and Psychiatry, 2*, 297–341.

Bebbington, P. E., Hurry, J., & Tennant, C. (1981). Psychiatric disorders in selected immigrant groups in Camberwell. *Social Psychiatry, 16*, 43–51.

Bhabha, Homi K. (1994). *The location of culture*. London: Routledge.

Billig, M. (1979). *Psychology, racism and fascism* (Birmingham: A. F. and R. Publications). Retrieved on July 26, 2016 from http://www.psychology.uoguelph.ca/faculty/winston/papers/billig/billig.html.

Bonilla-Silva, E. (2014). *Racism without racists. Color-blind racism ad the persistence of racial inequality in America* (4th ed.). New York: Rowman and Littlefield.

Bourne, J. (2001). The life and times of institutional racism. *Race and Class, 43*, 7–22.

Carmichael, S., & Hamilton, C. V. (1967). *Black power. The politics of liberation in America*. New York: Random House.

Carothers, J. C. (1947). A study of mental derangement in Africans and an attempt to explain its peculiarities, more especially in relation to the African attitude to life. *British Journal of Psychiatry, 93*(392), 548–597.

Carothers, J. C. (1951). Frontal lobe function and the African. *British Journal of Psychiatry, 97*(406), 12–48.

Carothers, J. C. (1953). *The African mind in health and disease: A study in ethnopsychiatry* WHO Monograph Series No. 17 (Geneva: World Health Organisation). Retrieved on July 26, 2016 from http://media.aphelis.net/wp-content/uploads/2013/10/WHO_MONO_17_part1.pdf.

Cohen, P. (1992). It's racism what dunnit'. Hidden narratives in theories of racism. In J. Donald & A. Rattansi (Eds.), *'Race', culture and difference* (pp. 62–103). London: Sage in association with Open University.

Cohen, Phil. (1999). *New ethnicities, old racisms?*. London: Zed Books.

Coon, C. S. (1963). *The origin of races*. London: Jonathan Cape.
Cooper, J. E., Kendell, R. E., Gurland, B. J., Sharpe, L., Copeland, J. R. M., & Simon, R. (1972). *Psychiatric Diagnosis in New York and London*, Maudsley Monograph No. 20. London: Oxford University Press.
Durrheim, K., Hook, D., & Riggs, D. W. (2009). Race and racism. In I. Prilleltensky, S. Austin, & D. Fox (Eds.), *Critical psychology. An introduction*. Los Angeles: Sage.
Easterly, W. (2006). *The white man's burden. Why the west's efforts to aid the rest have done so much ill and so little good*. Oxford: Oxford University Press.
Essed, P. (1990). *Everyday racism*, 2nd edn. C. Jaffé. Trans. (Alameda, CA: Hunter House). Originally published in Dutch as *Alledaags Racisme* (Baarn, The Netherlands: Ambo b.v.).
Eysenck, H. J. (1971). *Race, intelligence and education*. London: Temple Smith.
Eysenck, H. J. (1973). *The inequality of man*. London: Temple Smith.
Fanon, F. (1967). Racism and culture (text of Franz Fanon's speech before the first congress of Negro writers and artists in Paris, September 1965, and published in the special issue of *Présence Africaine*, June-November, 1956). In F. Maspero (Ed.) *Toward the African revolution. Political essays* (pp. 31–44) (transl. H. Chevalier), New York: Grove Press.
Fernando, S. (1988). *Race and Culture in Psychiatry*. London: Croom Helm. Reprinted as paperback Routledge, London 1989.
Garner, S. (2010). *Racisms; An introduction*. London: Sage.
Geary, D. (2015). *Beyond civil rights. The Moynihan report and its legacy*. Philadelphia: University of Pennsylvania Press.
Gibson, C., & Jung, K. (2012). *Historical census statistics on population totals by race, 1790 to 1990, and by hispanic origin, 1970 to 1990, for the United States, regions, divisions, and states*. Population Division Working Paper No. 56 (Washington, DC: U.S. Census Bureau).
Gilroy, P. (1987). *There ain't no black in the union jack. The cultural politics of race and nation*. London: Hutchinson.
Goldberg, D. T. (1993). *Racist culture philosophy and the politics of meaning*. Oxford: Wiley-Blackwell.
Hall, S. (1980). Race, articulation and societies structured in dominance. In UNESCO (ed.) *Sociological theories: Race and colonialism* (pp. 305–345). Paris: UNESCO. (Reprinted in Goldberg, D. T. and Essed, P. (eds). (2002). *Race critical theories* (pp. 38–68). Malden: Blackwell).
Hall, S. (1992). New ethnicities. In J. Donald & A. Ratansi (Eds.), *'Race', culture and difference* (pp. 252–259). London: Sage.
Hall, S., Critcher, C., Jefferson, T., Clarke, J., & Roberts, B. (1978). *Policing the crisis. Mugging, the state, and law and order*. Basingstoke: Macmillan.
Hepple, B. A. (1966). Race relations act 1965. *The Modern Law Review, 29*(3), 306–314.

Herrnstein, R. J., & Murray, C. (1994). *The Bell Curve: Intelligence and Class Structure in American Life*. New York: Free Press.

Home Department. (1999). *The Stephen Lawrence inquiry. Report of an inquiry by Sir William Macpherson of Cluny* Cm 4262-I (London: The Stationery Office). Retrieved on October 10, 2016 from https://www.gov.uk/government/publications/the-stephen-lawrence-inquiry.

Howe, D. (2011). New cross: The blaze we cannot forget. *The Guardian*, 17 January 2011. Retrieved on March 20, 2017 from https://www.theguardian.com/commentisfree/2011/jan/17/new-cross-fire-we-cant-forget.

Janson, M., & Saathoff, G. (eds.). (2009). *A mutual responsibility and a moral obligation: The final report on Germany's compensation programs for forced labor and other personal injuries*. Basingstoke: Palgrave Macmillan.

Jarvis, E. (1852). On the supposed increase of insanity. *American Journal of Insanity, 8,* 333–364.

Jensen, A. R. (1969). How much can we boost IQ and scholastic achievement? *Harvard Educational Review, 39,* 1–123.

Kamin, L. J. (1974). *The science and politics of IQ*. London: Wiley.

Kardiner, A., & Ovesey, L. (1951). *The mark of oppression. A psychosocial study of the American Negro* (Paperback ed.). New York: Norton.

Keller, R. C. (2007a). *Colonial madness. Psychiatry in French North Africa*. Chicago: University of Chicago Press.

Keller, R. C. (2007b). Taking science to the colonies: Psychiatric innovation in France and North Africa. In S. Mahone & M. Vaughan (Eds.), *Psychiatry and empire* (pp. 17–40). Basingstoke: Palgrave Macmillan.

Kraepelin, E. (1913). *Manic depressive insanity and paranoia*, trans. of *Lehrbuch der Psychiatrie* R. M. Barclay, 8th edn., vols. 3 and 4 (Edinburgh: Livingstone).

Kraepelin, E. (1920). Die Erscheinungsformen des Irreseins, *Zeitschrift fur die gesamte Neurologie and Psychiatrie,* Band 62, pp. 1–29, transl. H. Marshall; reprinted as 'Patterns of Mental Disorder' in S. Hirsch and M. Shepherd (Eds.) *Themes and Variations in European Psychiatry* (pp. 7–30). Bristol: John Wright 1974.

Kraepelin, E. (1921). *Manic depressive insanity and paranoia* (R. M. Barclay and G. M. Robertson Ed. and Trans). Edinburgh: Livingstone.

Lawrence, E. (1982). In the abundance of water the fool is thirsty: Sociology and black "pathology". In Centre for contemporary cultural studies (Ed) *The Empire Strikes Back: Race and Racism in 70s Britain* (pp. 95–142). London: Hutchinson.

Leff, J. (1973). Culture and the differentiation of emotional states. *British Journal of Psychiatry, 123,* 299–306.

Leff, J. (1977). The cross-cultural study of emotions. *Culture Medicine and Psychiatry, 1,* 317–350.

Lewin, M. (2009). Schizophrenia "epidemic" among African Caribbeans spurs prevention policy change. *Society Guardian* 9 December 2009. Retrieved on

July 30, 2016 from https://www.theguardian.com/society/2009/dec/09/african-caribbean-schizophrenia-policy.

Lewis, A. (1965). Chairman's opening remarks. In A. V. S. De Rueck & R. Porter (Eds.), *Transcultural psychiatry* (pp. 1–3). London: Churchill.

Mansoor, S. (2008). Black Britain's darkest hour. *The Guardian*, 24 February, 208. Available on 16 August at: http://www.theguardian.com/politics/2008/feb/24/race.

Maudsley, H. (1867). *The physiology and pathology of mind*. New York: D. Appleton.

Maudsley, H. (1879). *The pathology of mind*. London: Macmillan.

McCulloch, J. (1983). *Black Soul White Artefact. Fanon's Clinical Psychology and Social Theory*. Cambridge: Cambridge University Press.

Miles, R. (1984). The riots of 1958: Notes on the ideological construction of "race relations" as a political issue in Britain. *Immigrants and Minorities*, 3(3), 252–275.

Mishra, P. (2017). *Age of anger. A history of the present*. London: Allen Lane Penguin Random House.

Moon, C. (2009). *Narrating political reconciliation: South Africa's truth and reconciliation commission*. New York: Lexington Books.

Moynihan, D. (1965). *The Negro family in the United States: The case for national action*. Washington: US Governmental Printing Office.

Olusoga, D. (2016). *Black and British. A forgotten history*. London: Macmillan.

Omi, M., & Winant, H. (1994). *Racial formation in the United States* (2nd ed.). London: Routledge.

Omi, M., & Winant, H. (2015). *Racial formation in the United State* (3rd ed.). New York: Routledge.

Pilkington, E. (1988). *Beyond the mother country: West Indians and the Notting Hill white riots*. London: I. B. Taurus.

Prince, R. (1968). The changing picture of depressive syndromes in Africa. *Canadian Journal of African Studies*, 1, 177–192.

Pritchard, J. C. (1835). *A treatise on insanity and other disorders affecting the mind*. London: Sherwood, Gilbert and Piper.

Race Relations Act. (1968). London: Her Majesty's Stationery Office.

Race Relations Act. (1976). London: Her Majesty's Stationery Office.

Richards, G. (2012). *'Race', racism and psychology. Towards a reflexive history* (2nd ed.). London: Routledge.

Rosen, G. (1968). *Madness in society*. New York: Harper and Row.

Sashidharan, S. P. (1986). Ideology and politics in transcultural psychiatry. In J. L. Cox (Ed.), *Transcultural psychiatry* (pp. 158–178). London: Croom Helm.

Silverman, J. (1986). *Independent inquiry into the Handsworth disturbances*. Birmingham: Birmingham City Council.

Soanes, C., & Stevenson, A. (2008). *Concise Oxford English dictionary*. 11th edn Revised. Oxford: Oxford University Press.

Solomos, J., Findlay, B., Jones, S., & Gilroy, P. (1982). The organic crisis of British capitalism and race: the experience of the seventies. In Centre for Contemporary Studies (Ed.), *The Empire Strikes Back. Race and racism in 70s Britain*, (pp. 9–46). London: Hutchinson.

Stott, D. H. (1983). *Issues in the intelligence debate*. Windsor: NFER-Nelson Publishing.

Thomas, A., & Sillen, S. (1972). *Racism and psychiatry*. New York: Brunner/Mazel.

Tuke, D. H. (1858). Does civilization favour the generation of mental disease? *Journal of Mental Science, 4,* 94–110.

UN (United Nations). (1965). *United nations international Convention on the Elimination of All Forms of Racial Discrimination (CERD)* Geneva Office of the High Commissioner, United Nations Human Rights. Retrieved on February 10, 2017 from http://www.ohchr.org/EN/ProfessionalInterest/Pages/CERD.aspx.

UNESCO (United Nations Educational, Scientific and Cultural Organisation). (1950). *The race question*. Paris: UNESCO.

Warner, R. (1985). *Recovery from Schizophrenia. Psychiatry and political economy*. London, Boston and Henley: Routledge & Kegan Paul.

Watson, P. (1973). Race and intelligence through the looking glass. In P. Watson (Ed.), *Psychology and race* (pp. 360–376). Harmondsworth: Penguin.

Weigel, M. (2016). Political correctness: How the right invented a phantom enemy. *The Guardian*, 30 November 2016. Retrieved on December 10, 2016 from https://www.theguardian.com/us-news/2016/nov/30/political-correctness-how-the-right-invented-phantom-enemy-donald-trump.

Wellman, D. (1977). *Portraits of white racism*. Cambridge: Cambridge University Press.

World Health Organisation. (1973). *Report of the International Pilot Study of Schizophrenia* (Vol. 1). Geneva: WHO.

Yancey, W. L., Erickson, E. P., & Julian, R. N. (1976). Emergent ethnicity: A review and reformulation. *American Sociological Review, 41,* 391–402.

Zakaria, A. (2015). *Footprints of partition. Narratives of four generations of Pakistanis and Indians*. London: HarperCollins.

CHAPTER 5

Racism in a Context of Multiculturalism

As a result of higher levels of immigration from the 1950s onwards, the UK saw considerable change in its social structure in the fifty years that followed WWII. The UK is often said to have become multicultural, policies to encourage cultural diversity having been actively promoted in the public sector services, including the National Health Service (NHS), which provides mental health services. It is beyond the remit of this book to discuss details of what exactly the terminology surrounding 'culture' means but it is necessary to point out two issues that sometimes hamper discourse in matters to with racism. First, there is conflation often between the terms 'culture' and 'race', something that race theorists have subsumed within the notion of 'racialisation' (see section on 'Racialisation' later in this chapter). Second, it should be noted that the term ethnicity includes both notions of culture and of race, so the use of this term sometimes leads to a muddle about what exactly is meant. Also it is important to point to the interplay between discrimination, diagnosis and power whenever a society is composed of several social levels—especially when 'race' determines one of the borders.

5.1 Discrimination, Diagnosis and Power

The history of psychiatry shows that the need to diagnose mental illness has been intimately tied up with the need to control populations and people— the exercise of power over the individual often by, or on behalf of, the state (see 'Sociopolitical context' in Chap. 2). It was pointed out earlier (see 'Mental pathology and the construction of race-linked illnesses' in Chap. 3) that runaway black slaves were diagnosed as suffering from the illness 'drapetomania'—from '*drapetos*' a Greek word for a runaway slave and 'mania' representing a sort of compulsion—the main symptom being absconding from slavery. The use of diagnosis for control rather than for medical purposes was exemplified in the case of the so-called 'political abuse' of psychiatry in the former Soviet Union (Bloch and Reddaway 1984). In short, some political dissidents were sent to secure hospitals, especially the Serbsky Institute in Moscow and the Bekterev Institute in Leningrad, having been diagnosed as suffering from 'schizophrenia' because of their bizarre behaviour, delusional and grandiose ideas and other so-called 'symptoms'. During Stalinist times, psychiatry in the Soviet Union had a very low profile; it was with the liberalisation of the Soviet political system under Khrushchev that so-called 'abuse of psychiatry' occurred (Foucault 1988).

I recall visiting the Bekterev in the 1960s during a tourist visit to Leningrad and talking to some junior doctors there. My impression was that they had very hazy notions of the borderline between compulsory admission and informal (voluntary) admission of patients. Their approach was that if the doctor says someone needs admission then there cannot be any question as to the need for admission; and diagnosis was just a label used to regularise admission to hospital. My efforts to discuss the social impact of diagnosis really got nowhere in the short discussions (in limited English) that I had time for. It was likely that many of the diagnoses made when a troublemaking dissident came before a psychiatrist were perfectly respectable within the psychiatric medical model when applied in a context of a communist totalitarian system—at least the presence of illness within the system of psychiatry was used very naturally for confining someone who apparently needed to alter their behaviour or beliefs. For example, people who were sent to hospital were indeed showing what may have appeared to doctors as bizarre behaviour, irrational thought processes and so on, and could even have been seen as suffering from delusions if their beliefs were viewed in the narrow medical model of illness that prevailed in the context of the culture of the Soviet society at the time. In other words, what was referred to in the West as 'abuse' was in fact the *use* of psychiatric diagnosis (usually schizophrenia) as a tool for the control of people that the state wanted to exclude from society.

A similar situation to that in the Soviet Union occurred in apartheid South Africa. Apart from the (white) South African state practising apartheid in patient care and staff appointments, resulting in inequality of mental health care based on race, Smith Mitchell and Co., a white-owned company run for profit, provided custodial 'care' for 10,000–20,000 black patients, under contract to the government. A delegation from the American Psychiatric Association (APA) (Stone et al. 1979) visited South Africa in 1978 and reported on the 'enormous discrepancy between white and black facilities … [and that] the decision to transfer patients to Smith Mitchell facilities [was] predicated on the economic constraints dictated by apartheid' (p. 1505). What was called 'industrial therapy' in the Smith Mitchell Institutes included the maintenance of buildings and subcontracting of patient labour to other firms (Leading Article 1977). The shocking thing was that the (all-white) Society of Psychiatrists of South Africa (SPSA) made no attempt to counteract the racism of its psychiatric services. The SPSA even claimed that 'very extensive and advanced psychiatric services [are] given to all South Africans without reference to colour or creed' (Gillis 1977, pp. 920–921). It is easy to see how psychiatric diagnoses, carrying stigma and a sense of 'madness' (and the fear these engender) could be used socially for control, unless the professionals involved take pains to control the biases and prejudices that come along with the baggage of the disciplines they have been trained in. The campaign to persuade the Royal College of Psychiatrists (RCP) to put pressure on the SPSA to distance itself from apartheid is described in the section 'Action on apartheid' in Chap. 6.

5.2 Ethnic Issues in Mental Health Services

The inequalities in mental health system are summarised in Table 5.1. They reflect complex issues of race and culture interacting with those of class, gender, poverty and so on, but there is little doubt that racism plays a major role. I do not mean the overt racial prejudice of professionals involved in mental health systems but the racist attitudes embedded in the (psychiatric) diagnostic process: in the way assessments are carried out, in the colour-blindness involved in ignoring ways of thinking other than those of European culture and the insights derived from 'white knowledge', and so on (see Fernando 2010 and sections on 'How whiteness operates' and 'White knowledge' in Chap. 7). It should be noted that psychiatry is not an objective science: the empirical approach whereby psychiatrists identify mental 'phenomena' that constitute 'psychopathology' (such as 'guilt' 'self-depreciation', 'depression', 'feelings of passivity' etc.) implies that these descriptions get through to a reality that goes deeper than mere

Table 5.1 Racial inequalities in the UK

Black/Ethnic minorities more often:
- Diagnosed as schizophrenic
- Sectioned under Mental Health Act
- Admitted as 'offender patients'
- Held by police under MHA for observation
- Transferred to locked wards from open wards
- Not referred for 'talking therapies'

appearance; but these spurious 'facts' (for example, the presence or absence of psychopathology) are created under the influence of a socially constructed reality and culture (Harari 2001), and reflect the world views and values of the psychiatrist, together with the assumptions that underpin the 'culture' of psychiatry and its allied discipline, clinical psychology (see Fernando 2010).

One of the main issues in psychiatry and clinical psychology is the relatively high rate at which 'schizophrenia' is diagnosed among black people in the UK—often referred to as the 'over-representation' issue. In the 1980s, several studies found that black people in psychiatric hospitals in the UK were being labelled as 'dangerous' without an adequate objective reason for this (for example, Harrison et al. 1984) and black patients were over-represented among compulsorily detained patients in hospital (see Ineichen et al. 1984; McGovern and Cope 1987). This situation has not changed. It is likely that similar situations are present other European countries. In the USA too, there is an over-representation of African Americans amongst involuntary commitments to public mental institutes (Lindsey and Paul 1989; Lawson et al. 1994).

The nature of embedded racism resulting in schizophrenia becoming a 'black disease' in the USA has been explored by Jonathan Metzl (2009), and a similar process in the UK has been explored by me in a recent book chapter (Fernando 2017). The issue of racial bias in diagnostic style (which in effect amounts to racism) is not easy to examine but was the topic of an exceptional piece of research by Loring and Powell (1988) which used carefully constructed vignettes (of case histories). When the vignettes were sent out to 290 black and white American psychiatrists with a series of questions, the researchers found that: (a) overall, black clients, compared to white clients, were given a diagnosis of schizophrenia more frequently by both black and white clinicians—although this was done to a lesser extent by the former; and (b) all the clinicians appeared to ascribe violence, suspiciousness and dangerousness to black clients even though the case studies were the same as those for the white clients. Loring and Powell (1988) concluded that black and white people are 'seen differentially even if they exhibit the same behaviour', and that 'these differences will be reflected and legitimized in official statistics on psychopathology' (p. 19). The lessons are clear: unless we look at institutional processes, we are unlikely to find out why and how diagnoses reflect and perpetuate racism.

Reliable ethnic statistics on the use of psychotherapy/counselling services (usually overseen in the UK by clinical psychologists) are non-existent, although anecdotal evidence is that people from black and Asian communities are seldom referred for such therapy in the NHS (Campling 1989). A high 'drop-out' rate from psychotherapy was reported many years ago in the case of black patients in the USA (for example, see Rosenthal and Frank 1958; Yamamoto et al. 1967, 1968) and it is likely that a similar situation exists in Britain too (personal observation). The question is not just about access to services providing 'psychotherapy' or 'counselling' but more importantly the nature of what is offered by those services. One needs to consider the appropriateness of the therapy in terms of models used in understanding (or helping the clients to understand) the problems presented; attitudes and ideologies of the people who provide 'therapy'; and so on. The overall impression is that many services reflect racist attitudes and racist deductions are made there about clients. In my experience, many black people are disappointed with psychotherapy and counselling provided at generic centres, whether private, voluntary or statutory. However, I know from talking to people working at ethnic-specific counselling centres (i.e. those that aim to provide counselling for specific ethnic groups) that clients attending them are often highly satisfied with the services they receive (see Fernando 2005).

5.3 Racialisation

Omi and Winant (2015) write: 'Through a complex process of selection, human physical characteristics ("real" or imagined) become the basis to justify or reinforce social differentiation. Conscious or unconscious, deeply ingrained or reinvented, the making of race, the "othering" of social groups by means of the invocation of physical distinctions, is a key component of modern societies' (p. 111). The notion of 'race' came about as a result of a historical process (see 'Exploration, colonialism, race-slavery' in Chap. 2) and its definition is discussed elsewhere in this book (see 'Definitions of racism and race' in Chap. 4). Although it was originally defined by specific features of the human body (mainly skin colour)—commonly thought of as being a biologic marker(s) for inferiority—it is now a sociopolitical concept defined by particular experiences undergone by people. However, because of its history, the definition holds on to a reference to *types of human bodies* and because of its social connotations the definition incorporates the lived experiences (journeys) of individual people—people with those bodies (see Omi and Winant's social definition in 'Definitions of racisms and race' in Chap. 4). Although racialisation broadens the range of 'bodies', it is necessary for those bodies to be 'different', different that is to the 'white' body.

The notion of 'racialisation' was first introduced by Frantz Fanon as indicative of dehumanisation of various groups of people by structuring them as 'races' in the manner that Europeans identified people subjugated (by white people) during slavery and colonialism (see Fanon 1967). But the term racialisation is now used more widely and less specifically and not necessarily for the original formation of the notion of 'race': it speaks to how discrimination and oppression is enacted through groups of people (for example culturally defined groups, or people defined by nominal religious affiliation) being seen *as if they are races*, referring to the original classifications of races that were determined on an assumption of fundamental biological differences between them. In effect, the notion of 'race' is applied to (say) cultures, religions or kinds of people that are thought of in the same way as 'races'—groups that are discriminated against and usually recognisable not just by physical appearance, *by their bodies*, but by a host of other factors and so do not necessarily have to be that different in skin colour from the 'white'. Thus, groups of people who could pass for 'white' such as Parsis and some people from Middle Eastern (West Asian) countries, have been racialised in the British context—they are seen as a racial group because of their heritage. The process of racialisation has resulted in categories of people who may not usually be seen as 'black' or 'Asian' ('brown-skinned') being included in the British category 'black and minority ethnic' or BME (sometimes slightly lengthened to 'black, Asian and minority ethnic' (BAME). Those included are people who are *racialised*.

In some settings we find people seen as belonging to 'alien cultures' being racialised—for example when in 1978 Margaret Thatcher talked of Britain being 'swamped by people with a different culture' (Fitzpatrick 1990, p. 249)—a line of argument clearly expressing racist sentiments—she was racialising immigrants 'with a different culture', and during the campaigning in the months leading up to the 2016 referendum on leaving the European Union, similar descriptions of the racial 'Other' were used (see also 'New era of unashamed racism?' in Chap. 9). Racialisation is sometimes taken further —if we use the term more loosely—to refer not just to persons (their bodies) but to social practices, such as the process of diagnosis in psychiatry and psychology. Thus schizophrenia is said to be racialised in some settings by the diagnosis being given predominantly to black people (see 'Racialisation of the schizophrenia diagnosis' later in this chapter).

A term that alludes to some features of racialisation is the term 'politically black'. When the Transcultural Psychiatry Society (TCPS) was active in the UK in the 1980s (see 'Transcultural psychiatry in the UK' in Chap. 6), there was intense political activity by the Irish Republican Army (IRA) and its sympathisers, who demanded that Northern Ireland (part of the UK) should be split off from the UK and join the Irish Republic to form one Irish state. I recall that the TCPS had some members from an Irish background, who sympathised with the anti-racist stance of the TCPS. Essentially the Irish members of the TCPS claimed to be 'politically black', as they were from a country that had suffered from British colonialism and (in their view) they suffered from discrimination and other disadvantages in the way that 'black people' suffered in the UK—in other words, they had the 'black experience'. I recall that when I first came to the UK as a student, I saw window signs in some parts of London reading 'No coloureds, no Irish' meaning that rooms were available for letting, but not for certain groups— these would have been racialised groups at that time. However, as time went by, the reluctance of landlords to let rooms to Iris people diminished and one could then envisage that the Irish had been deracialised. This line of thinking has been pursued in books by Ignatief (1995) and Brodkin (2010) where the special historical position of Jews and Irish in the USA are discussed in terms of being seen as (politically) 'black' in a certain context but then later being seen as 'white'. If we accept the legitimacy of 'political blackness', we need to consider 'political whiteness'. 'Passing for white' is a well-known phenomenon whereby people generally thought of as black people deny their blackness for some reason, but usually only when they can get away with it by being light-skinned; however there is the well-known instance of a white woman, Rachel Dolezal, who was a respected black rights worker until she was outed as being 'white' and who now claims to have chosen a black identity (Aitkenhead 2017).

Since the appearance in the UK of new racisms (see 'New racisms in the UK' in Chap. 4), notions of 'race' seem to have become varied and complex and, more recently, complicated by dubious claims to being either black or white, such as those of Dolezal. The best we can do is to stay as close as possible to the social definition of race offered by Omi and Winant (2015) and discussed in 'Definitions of racism and race' in Chap. 4.

5.4 Racism in 'Psy' Research

There are many fundamental problems with psychiatric and psychological research ('psy' research), which is carried out in a context of institutional racism. I shall summarise them briefly. The first problem arises at the point of planning the questions to be asked. 'Psy' research, more than research in any other field of medicine or psychology, is politically driven. Apart from the obvious political forces like the pharmaceutical industry (that often funds psychiatric research), there are others (sometimes difficult to identify easily) that determine what is researched and what is not researched. In the field of clinical psychology, it is much more common to research matters that emanate from Western knowledge ('white knowledge')—such as, say, depressive feelings or related emotional states, than it is to research, say, spirituality or filial piety, which are associated with Indian or Chinese traditions. And, what is a cultural bias resembles (if not identical with) racial bias. The underlying agenda is that white knowledge supersedes any other knowledge (see 'White knowledge' in Chap. 7). But the greater issue is around how research is carried out.

The research strategy seldom, if ever, allows for the need to compensate for the inherent bias of researchers as revealed, for example, in the report *Understanding Psychosis and Schizophrenia* (see 'Racism of a psychology report' in Chap. 8). Even more importantly, research methods tend to be blind to the serious drawbacks of relying on empiricism as a scientific foundation for research in psychiatry (Harari 2001). Most psychiatric research starts off by marshalling so-called observable measurable data, assuming that all the data are value-free—or at least reasonably 'objective'—and assumes that what is measured (and what is left out) is chosen at random or more often, because of clear-cut reasons. The fact is that the selection (made by choice) could be a hotbed for racist agendas or at least bias arising from institutional attitudes in psychology or psychiatry. Items measured as 'phenomena' (such as depression, thought disorder and hallucinations) are not objective 'facts' (in the case of 'psy' disciplines) although often treated as such. They are essentially judgements made by people trained in a particular cultural framework and so the conclusions drawn are value-laden and theory-laden (see Harari 2001). This is firm ground for institutional racism. Further, psychological and psychiatric data —unlike those in 'hard' natural sciences—are similar to data in social psychology in that they (the data) deal in events that fluctuate markedly over time, but are presented as static 'facts' (Gergen 1973)—again, racism can easily creep in at this level. For example, a propensity to aggressiveness, feelings of depression or 'hearing voices' may be noted as relatively stable 'observations' about a person and conclusions may be drawn about them— but only if the contextual realities around the events that were construed (as aggressiveness, feelings of depression, hallucinations) are ignored. In all of these cases, when judgements are being made in an essentially racist society, the context for white people may be very different to that for racialised people (see 'Sociopolitical context' in Chap. 2 for discussion of how the 'psy' disciplines came to incorporate racist notions in their structures).

In the case of cross-cultural research, especially 'epidemiological' research (measuring incidence or prevalence of illnesses as defined within Western psychiatry), the number of locations where racism can be active is immense. The main issue is around the question of validity of diagnoses— their usefulness when used across cultures. Kleinman (1977) has called this the problem of 'category fallacy' (p. 4): for example in the case of diagnosing depression, the depressive syndrome (that the 'psy' disciplines recognise) 'only represents a small fraction of the entire field of depressive phenomena', which differ across cultures. Therefore applying such a category to analyse cross-cultural studies would result in 'systematically missing what does not fit its tight parameters' (pp. 3–4). Kleinman's arguments apply even more cogently to diagnoses such as schizophrenia. In many research studies, 'race' and 'culture' are conflated, and often racism intrudes by influencing the conclusions that are arrived at. The basic problem is that the 'psy' disciplines, having developed within a specific tradition—broadly termed 'Western culture' (see Chap. 2)—the particular illness model (with its diagnoses) that is applied reflects the world view about the human condition within that tradition. The situation vis-à-vis a different cultural tradition may be very different and this is where racism comes into the picture. What is 'found' as a cultural difference gets (mis) identified as something racial. One could argue that if a system—the psychiatric system in this case—is sufficiently flexible it could take on changes in world view to suit each particular situation, but this is not the case. The basic syndromes agreed as 'illnesses' are adhered to as unchanging and, even more importantly, the main symptoms recognised are perceived as 'phenomena' that have an existence of their own. Kleinman (1977) suggests that an ideal cross-cultural study should begin with phenomenological descriptions that are indigenous to each cultural group. He believes that researchers may then 'elicit and compare symptom terms and illness labels independent of a unified framework' (p. 4).

5.5 Manipulation of Research Findings

In the 1980s and 1990s, psychiatric research into what are called 'ethnic issues' were popular, while services for black and Asian communities seemed to get worse. The array of 'research findings' on ethnic issues were mainly concerned with counting numbers of people given various diagnoses in what are termed 'epidemiological' surveys. One of the main issues focused on was the over-representation of black people amongst those patients deemed to be 'schizophrenic' (something that is still evident). Successive surveys made similar statistical conclusions but did not really advance knowledge about the reason for this 'over-representation' or come up with any useful suggestions on improving clinical practice. Apart from a study by Parkman et al. (1997) which was inadequate in that it merely touched the surface of a vast problem, there have been no studies (that I know of) published in professional journals that have drawn on the experience of black patients (from their viewpoints) or their views about diagnosis and treatment. Nor have there been any studies of racism in the psychiatric system except for a study by Lewis et al. (1990), which attempted to examine racial bias in diagnosis. Unfortunately this study was seriously flawed because the information in the vignettes, which were supposed to be race-neutral, actually revealed the ethnicity of the clients—an indication of the researchers' lack of sophistication in handling the issue of 'race' (Fernando 1991).

Kuller (1999) has pointed out that good epidemiological studies 'progress from descriptive to analytic to experimental epidemiology and then to studies of effectiveness leading to prevention' (p. 897). Eaton and Harrison (2000) have pointed out that the lack of progress in British epidemiological psychiatric research into ethnic issues resembles what Kuller terms 'circular epidemiology', which is the continuation of a particular line of research after something has been established beyond reasonable doubt. This is likely to have occurred because researchers are unwilling or unable to question their methodology or test out hypotheses that may question dogmatically held ideologies—and the notion that 'research' is inherently race-free and culture-free is one. Further, the politics of research and getting access to reputable journals militates against challenging institutional racism. As Kuller (1999) puts it: '[A] new hypothesis for which there is lack of substantial prior data is unlikely to be successful in terms of peer review' (p. 897)—and hence research based on such a hypothesis is unlikely to attract funding. Epidemiological research into ethnic issues, carried out with little attention to the possibility of racial bias in its methodology, was—and is—stuck in a rut because researchers cannot or will not question the dogma of traditional diagnostic categorisation. Far too often, data on mental health problems (that may itself may be biased) is analysed in a framework of a narrow biomedical model that reduces complex human problems to simplistic and culturally insensitive definitions of 'illness' and 'normality' formulated within Western cultural traditions—and this amounts to institutional racism. The question that the 'psy' disciplines must face is about the usefulness of much of what goes on as 'research', especially research that claims to be 'epidemiological', and what the real (hidden) agendas may be. In the following two paragraphs I shall present two true stories and let the readers draw conclusions about hidden agendas.

A black psychiatrist from Jamaica was invited to work with a research team at a prestigious institute in London, in the mid-1990s on a study to investigate the diagnosis of schizophrenia among black people. The aim of the study was to explore discrepancies if any between the diagnoses made by the Jamaican psychiatrist and those made by native (white) British psychiatrists. When the black Jamaican colleague, by then resident abroad, saw a draft of the paper written by the others involved in the research (but which included his name as one of the authors) he found that (in his view) the data, which seemed to show that there was no significant difference in diagnostic practice between that of the Jamaican practitioner and the native British psychiatrists, was incomplete. In his view, if the full data set were analysed it would show that there *were* significant differences. He wanted the paper changed, only to be told that it had already been peer-reviewed and accepted for publication and so could not be changed. When the Jamaican psychiatrist threatened legal action to prevent the publication using (what he considered to be) incorrect statistics, the paper was withdrawn and re-written.

The second story is about getting funding for research from the research committee connected with special hospitals (hospitals with maximum security). In 1997 I was involved in applying for funding for a project to survey and evaluate the views of black service users who had been patients in a secure hospital. In spite of my obtaining the support of the hospital staff concerned (which was not easy) and the support of black staff within the body that dealt with the application, the proposal was turned down on the basis of reports from the peer reviewers. When I enquired about the reasons for the rejection, I found that, although the project was about service users' views, no service user had been asked to review the project. Worse still there were neither service users nor black people on the panel that made decisions on the allocation of funds for research projects.

These stories—and the attitudes of 'psy' professionals alluded to in some other parts of this book (see for example the sections 'Racism of a psychology report' and 'Racist conclusions of psychiatric research' in Chap. 8) illustrate the political implications of 'psy' research and how easily institutional racism can permeate official reports written by professionals. The basic problem may be that the research (and publication) field today is highly competitive and often driven by economic gain for vested interests. Hence, if there are issues around race that appear to challenge the 'psy' disciplines, the interests of prestigious university departments or the status of particular individuals, the question of publication is fraught, partly because of the peer-review system in which the peers are usually establishment figures—and the same can be said about getting funds for research and the backing of universities.

The tragedy to my mind is that black people—researchers particularly—are increasingly inveigled into being involved in supporting dubious research through institutional racism. For example, black professionals are sometimes *used* (in the sense of taken advantage of) by being invited to be a part of the team, while at the same time being excluded from having any influence on how the research is conducted and the conclusions arrived at. When several years ago I protested to a civil servant at the Department of Health who was involved with the allocation of funding for what I and other black professionals considered to be a flawed research project into 'psychosis' among black people, I was told that it was 'alright' because a black registrar (junior doctor) was involved. I knew that this particular person needed to show research experience on her curriculum vitae to progress in her career and in fact was only marginally involved in the research project. In the mid-1990s, another black professional and I were invited to serve on an advisory committee at the Department of Health in order to review a research project into deliberate self-harm among Asian women. The project put up for funding by the Department of Health (DOH) envisaged assessments for clinical depression that were clearly inappropriate. When we (the two black people consulted on its suitability) wanted the project changed so that social and cultural dimensions were included alongside the medical approach proposed, we were both dropped from the committee reviewing the project and the research went ahead unchanged.

The main problem with psychiatric research in the field of ethnicity is that a complex field is being researched using telescopes that examine minutiae often determined by what researchers hope to find; what are seen as big issues, such as the influence of racism, are put to one side, too 'big' to research. In scientific terms, hard data which are useful in practice are very difficult to come by because the methods of research into ethnic issues are seriously flawed, essentially being too narrow-minded and reductionist. The fundamental problems of cross-cultural research, including the effect of racism in influencing the judgements that go to make diagnoses, have not been addressed. Further, the tools used to measure 'culture', 'racism' and allied matters are far too narrow to yield useful information; in fact the 'information' that sometimes emerges when tools that are inappropriate to the task are used is misleading. In simple terms, the research methodology in the 'psy' disciplines that is applied to matters concerned with cultural or racial difference often amounts to being institutionally racist. Moreover, the need to research institutional racism as such is not being tackled, possibly for political reasons. Notwithstanding the difficulties and problems, research into issues of race and culture ('ethnic issues') in the mental health field is necessary; but the tools must change and the agenda needs to be clarified if institutional racism is to be avoided.

5.6 Explanations for 'Schizophrenia' in Black People

When the high rates at which British-born black people are given the 'schizophrenia' label—'the over-representation issue'—first became evident in the 1980s, the medical establishment and the media took it up in the context of discussions about the black presence in Britain. An eminent psychiatrist from the Institute of Psychiatry quoted the so-called high 'incidence' of schizophrenia among black people as evidence that 'something had gone dreadfully wrong for the Caribbean second generation' (Ballantyne 1988, p. 6). An immunological/virus theory was proposed as an explanation for the over-representation issue. This theory was first proposed in a paper in the *British Journal of Psychiatry* (King and Cooper 1989) and was keenly taken up by researchers at the IOP such as Wing (1989), Harrison (1990), Eagles (1991), and Wessely et al. (1991). The alacrity with which institutional psychiatry, represented by eminent British psychiatrists, grabbed at the virus theory was most alarming considering that the evidence for the theory was almost non-existent; it was totally discredited a few years later (for example by Crow and Done 1992 and Cannon et al. 1996).

A professor of psychiatry from Nottingham University used the virus theory to explain the over-representation issue in a BBC television programme—a programme that was actually titled *Black Schizophrenia*—broadcast on 13 March 1989 and described briefly below. An alleged connection between 'black schizophrenia', and violence and the dangers of infections being imported into the UK by black immigration, was made obvious visually in the film by (a) statistics quoting high rates of schizophrenia among black people being discussed against images of rioting black youth; and (b) the virus theory being discussed against images of West Indians coming off boats. The racist undertones of the film were obvious. The programme was objected to by Mind (the National Association for Mental Health) and the Transcultural Psychiatry Society (TCPS) in a joint letter to the Director of the BBC, and there was correspondence (Fernando 1989; Brightwell 1989) in the official BBC journal *The Listener* about the programme, as well as a supporting article (Tilby 1989) (see section on 'Transcultural psychiatry in the UK' in Chap. 6 for further information on the TCPS).

In spite of many epidemiological reports, no other medical theories apart from the virus theory have been produced to explain the over-representation of black people being diagnosed as schizophrenic. However, the AESOP study (Fearon et al. 2006) claims to have found that a variety of social factors in the backgrounds of (black) people who were given the diagnosis of schizophrenia (see section on 'Racist conclusions of psychiatric research in Chap. 8); and researchers in the Netherlands (Selten and Canto-Graae 2005) have postulated that 'the long term experience of social defeat, defined as a subordinate position or as "outsider status"' (p. 101) may be a risk factor in schizophrenia. The explanation for the over-representation issue is complex, much more to do with the way diagnoses are made and the racialisation of the schizophrenia diagnosis in Western countries (see next section in the chapter), compounded by issues such as the added stress on racialised groups in British society because of multiple pressures and discriminations resulting from racism (for further discussion on this matter see Fernando 2010).

5.7 Racialisation of the Schizophrenia Diagnosis

The social construction of psychiatric diagnoses, including schizophrenia, was discussed briefly in Chap. 2. Naturally these diagnoses carry their own particular associations which may connect up with other ideas derived from, say, tradition, public media or just 'common sense'. In the context of 1980s Britain, where the public consciousness associated 'race' with drug abuse and the anger of black youth was attributed to their use of cannabis (Imlah 1985), the diagnosis of 'cannabis psychosis' became popular and was given almost exclusively to black African-Caribbean people (McGovern and Cope 1987). In addition to pressures arising from the context in which diagnoses are made, the diagnostic process is affected by many factors totally outside the medical or psychological domains during the recognition and evaluation of symptoms of psychopathology—and stereotypes play a big part here. American stereotypes of the patient who is perceived as 'non-Western', usually on the basis of colour, are described by Sabshin et al. (1970) as being that they are '… hostile and not motivated for treatment, having primitive character structure, not psycho-logically minded, and impulse-ridden' (1970, p. 788). Similar stereotypes prevail in Britain, derived from its colonial past: the influence of the 'big, black and dangerous' stereotype in determining diagnosis and treatment in British forensic psychiatry was highlighted in a report of an inquiry into the deaths of three black men in a forensic hospital in the early 1990s (SHSA 1993); and the perceived dangerousness of black people seems to lead to the excessive use of seclusion or high levels of medication when they are diagnosed as 'schizophrenic'.

Post-Enlightenment thinking in the 'psy' disciplines as they developed in Europe during the nineteenth century was strongly influenced by two main concepts—'degeneration' as a basis for understanding poverty, lunacy and racial inferiority; and the idea of the 'born criminal' derived from the scientific study of crime. The former was dominated by Eugène Morel (1852, 1857), and the latter by Cesare Lombroso (1871, 1911) who focused on studying the apparent 'racial' variation in the nature of human beings across Italy from north to south. Morel internalised degenerationas representing lesions in the brain or the mind, but Lombroso, who believed firmly that the white races represented the triumph of the human species, felt that 'inside the triumphant whiteness, there remained a certain blackness' (Pick 1989, p. 26), implying that degeneration could be located in the 'Other'—the criminal, the insane and a hierarchy of races that harboured 'blackness'. In this context, German psychiatrist Emil Kraepelin (1896) constructed a specific illness category, dementia praecox, which he saw as a biologically determined illness generated by organic lesions or faulty metabolism (Metzl 2009). The same condition was interpreted by Swiss psychiatrist Paul Eugen Bleuler (1911) as representing a loosening of associations in the mind, due to a split (*schizo*) in the psyche (*phrene*), and Bleuler therefore renamed the illness 'schizophrenia', essentially a disorder of personality. Jonathan Metzl (2009), in reviewing the adoption of these ideas in the USA, reckons that when they crossed the Atlantic during the Jim Crow era (see the section 'Exploration, colonialism, race-slavery' in Chap. 2), the Kraepelinian version (of dementia praecox/schizophrenia) was attached to African Americans deemed 'mentally ill' (and confined to asylums) because it tallied with the image of black Americans as marginalised people (Stonequist 1937), while schizophrenia in white people was seen as a disease of sensitive people suffering from psychological trauma (and not always confined to asylums). Metzl (2009) argues from the findings of his own research in Michigan that this differential use of the schizophrenia diagnosis became much more prominent in the USA following the civil rights movement: 'in addition to the diagnosis schizophrenia becoming racialized it also became a complex metaphor for race' (p. 109). In the UK too the schizophrenia diagnosis has been racialised to a large extent in that the 'race' is the main socially constructed determinant in the diagnosis when it is used by the 'psy' disciplines.

References

Aitkenhead, D. (2017). Rachel Dolezal: "I am not going to stoop and apologise and grovel". In *The Guardian*. 25 February, 2017. Retrieved April 7, 2017 from https://www.theguardian.com/us-news/2017/feb/25/rachel-dolezal-not-going-stoop-apologise-grovel.

Article, Leading. (1977). Apartheid and mental health care. *Lancet, 2,* 491.
Ballantyne, A. (1988, October 31). Young blacks vulnerable to schizophrenia, *The Guardian,* p. 6.
Bleuler, E. (1911). *Dementia Præcox or the Group of Schizophrenias* (J. Zitkin, Trans.). New York: International Universities Press (Reproduced 1950).
Bloch, S., & Reddaway, P. (1984). *The shadows over world psychiatry.* London: Gollancz.
Brightwell, R. (1989, April, 13). Letter in *The Listener,* p. 18.
Brodkin, K. (2010). *How the Jews became White folks and what that says about race in America.* New Brunswick, NJ: Rutgers University Press.
Campling, P. (1989). Race, culture and psychotherapy. *Psychiatric Bulletin, 13,* 550–551.
Cannon, M., Cotter, D., Coffey, V. P., Sham, P. C., Takei, N., Larkin, C., et al. (1996). Prenatal exposure to the 1957 influenza epidemic and adult schizophrenia: A follow-up study. *British Journal of Psychiatry, 168,* 368–371.
Crow, T. J., & Done, D. J. (1992). Prenatal exposure to influenza does not cause schizophrenia. *British Journal of Psychiatry, 161,* 390–393.
Eagles, J. M. (1991). The relationship between schizophrenia and immigration. Are there alternatives to psychosocial hypotheses? *British Journal of Psychiatry, 159,* 783–789.
Eaton, W. W. and Harrison, G. (2000) Epidemiology, social deprivation, and community psychiatry. *Current Opinion in Psychiatry, 13,* 185–7.
Fanon, F. (1967). Racism and culture (text of Franz Fanon's speech before the first congress of Negro writers and artists in Paris, September 1965, and published in the special issue of *Présence Africaine,* June–November, 1956). In F. Maspero (Ed.), *Toward the African revolution. political essays* (H. Chevalier Trans.), pp. 31–44. New York: Grove Press.
Fearon, P., Kirkbride, J. B., Morgan, C., Dazzan, P., Morgan, K., Lloyd, T., et al. (2006). Incidence of schizophrenia and other psychosis in ethnic minority groups: Results from the MRC AESOP Study. *Psychological Medicine, 26,* 1541–1550.
Fernando, S. (1988). *Race and Culture in Psychiatry.* London: Croom Helm. Reprinted as paperback Routledge, London 1989.
Fernando, S. (1989, March 30). 'Personality clash', Letter in *The Listener,* p. 19.
Fernando, S. (1991). Racial stereotypes. *British Journal of Psychiatry, 158,* 289–290.
Fernando, S. (2005). Multicultural mental health services: Projects for minority ethnic communities in England. *Transcultural Psychiatry, 42*(3), 420–436.
Fernando, S. (2010). *Mental health, race and culture* (3rd ed.). Basingstoke: Palgrave.
Fernando, S. (2017). 'Racialization of the Schizophrenia Diagnosis' In B. Cohen (Ed.), *Routledge international handbook of critical mental health.* New York: Routledge, (in press).

Fitzpatrick, P. (1990). Racism and the innocence of law. In D. T. Goldberg (Ed.), *Anatomy of Racism* (pp. 247–262). Minneapolis: University of Minnesota Press.

Foucault, M. (1988). *Politics Philosophy Culture. Interviews and Other writings 1977-1984.* In L. D. Kritzman (Ed.). London: Routledge.

Gergen, K. J. (1973). Social psychology as history. *Journal of Personality and Social Psychology, 26,* 309–320.

Gillis, L. S. (1977). Letter to the editor. *Lancet, 2,* 920–921.

Harari, E. (2001). Whose evidence? Lessons from the philosophy of science and the epistemology of medicine. *Australian and New Zealand Journal of Psychiatry, 35,* 724–730.

Harrison, G. (1990). Searching for the causes of schizophrenia; The role of migrant studies. *Schizophrenia Bulletin, 16,* 663–671.

Harrison, G., Ineichen, B., Smith, J., & Morgan, H. G. (1984). Psychiatric hospital admissions in Bristol II. Social and clinical aspects of compulsory admission. *British Journal of Psychiatry, 145,* 605–611.

Ignatief, N. (1995). *How the Irish became White.* New York: Routledge.

Imlah, N. (1985). *Silverman enquiry on handsworth riots* (Unpublished transcript cited in Fernando, 1988).

Ineichen, B., Harrison, G., & Morgan, H. G. (1984). Psychiatric hospital admissions in Bristol. I. Geographical and ethnic factors. *British Journal of Psychiatry, 145,* 600–604.

King, D. J., & Cooper, S. J. (1989). Viruses, immunity and mental disorder. *British Journal of Psychiatry, 154,* 1–7.

Kleinman, A. (1977). Depression, somatization and the "new cross-cultural psychiatry", *Social Science and Medicine, 11,* 3–10.

Kraepelin, E. (1896). *Psychiatrie* (5th ed.). Leipzig: Verlagvan Johann Ambrosius Barth.

Kuller, L. H. (1999). Invited commentary: Circular epidemiology. *American Journal of Epidemiology, 150*(9), 897–903.

Lawson, W. B., Hepler, N., Hollady, J., & Cuffal, B. (1994). Ethnicity as a factor in inpatient and outpatient admissions and diagnosis. *Hospital & Community Psychiatry, 45,* 72–74.

Lewis, G., Croft-Jeffreys, C., & David, A. (1990). Are British psychiatrists racist? *British Journal of Psychiatry, 157,* 410–415.

Lindsey, K. P., & Paul, G. L. (1989). Involuntary commitments to public mental institutions: Issues involving the overrepresentation of Blacks and assessment of relevant functioning. *Psychological Bulletin, 106*(2), 171–183.

Lombroso, C. (1871). *L'uomo bianco e l'uomo di colore. Letture sull'origine e le varietà delle razze umane* [White Man and the Coloured Man: Observations on the Origin and Variety of the Human Race] (Padua). Cited by Pick, 1989.

Lombroso, C. (1911). *Crime: Its causes and remedies* (H. P. Horton, Trans.). London: Heinemann.
Loring, M., & Powell, B. (1988). Gender, race and DSM-III: A study of the objectivity of psychiatric diagnostic behavior. *Journal of Health and Social Behavior, 29,* 1–22.
McGovern, D., & Cope, R. (1987). The compulsory detention of males of different ethnic groups, with special reference to offender patients. *British Journal of Psychiatry, 150,* 505–512.
Metzl, J. (2009). *The protest psychosis how schizophrenia became a black disease.* Boston, MA: Beacon Press.
Morel, B.-A. (1852). *Traites des Maladies Mentales* Cited by Metzl, 2009, p. 27. Paris: Masson.
Morel, B.-A. (1857). *Traite Des Degenerescences Physiques, Intellectuelles Et Morales De L'Espece Humaine Classics in Psychiatry Series Print-to-Order.* Delhi: Gyan Books.
Omi, M., & Winant, H. (2015). *Racial formation in the United State* (3rd ed.). New York: Routledge.
Parkman, S., Davies, S., Leese, M., Phelan, M., & Thornicroft, G. (1997). Ethnic differences in satisfaction with mental health services among representative people with psychosis in South London: PriSM study 4. *British Journal of Psychiatry, 171,* 260–264.
Pick, D. (1989). *Faces of degeneration. A European disorder, c. 1848–c. 1918.* Cambridge: Cambridge University Press.
Rosenthal, D., & Frank, J. D. (1958). The fate of psychiatric clinic outpatients assigned to psychotherapy. *Journal of Nervous and Mental Disease, 127,* 330–343.
Sabshin, M., Diesenhaus, H., & Wilkerson, R. (1970). Dimensions of institutional racism in psychiatry. *American Journal of Psychiatry, 127*(6), 787–793.
Selten, J.-P., & Cantor-Graae, E. (2005). Social defeat: Risk factor for schizophrenia. *British Journal of Psychiatry, 187,* 101–102.
SHSA (Special Hospitals Service Authority). (1993). *Report of the Committee of Inquiry into the Death in Broadmoor Hospital of Orville Blackwood and a Review of the Deaths of Two Other Afro-Caribbean Patients: 'Big, Black and Dangerous?' (chairman Professor H Prins).* London: SHSA.
Stone, A., Pinderhughes, C., Spurlock, J., & Weinberg, M. D. (1979). Report of the committee to visit South Africa. *American Journal of Psychiatry, 136,* 1498–1506.
Stonequist, E. V. (1937). *The marginal man.* New York: Scribner.
Tilby, A. (1989, March 16). Personality clash, *The Listener,* pp. 10–11.
Wessely, S., Castle, D., Der, G. and Murray, R. M. (1991). Schizophrenia and Afro-Caribbeans. A case-control study. *British Journal of Psychiatry, 159,* 795–801.

Wing, J. K. (1989). Schizophrenic psychoses: Causal factors and risks. In P. Williams, G. Wilkinson, & K. Rawnsley (Eds.), *The scope of epidemiological psychiatry* (pp. 225–239). London: Routledge.

Yamamoto, J., James, Q. C., Bloombaum, M., & Hattem, J. (1967). Racial factors in patient selection. *American Journal of Psychiatry, 124,* 630–636.

Yamamoto, J., James, Q. C., & Palley, N. (1968). Cultural problems in psychiatric therapy. *Archives of General Psychiatry, 19,* 45–49.

CHAPTER 6

Struggle Against Racism in the UK

In 1993, a young black student called Stephen Lawrence was murdered by a gang of white youth in South London. In spite of public protest at the failure of the London Metropolitan Police to investigate this murder properly, the Tory government at the time refused to hold an inquiry. However, after a change of government in 1997, the incoming Labour government instituted a public inquiry led by a retired judge, Sir William Macpherson. The publication of the report he produced—the Macpherson report (Home Department 1999)—was perhaps the most significant event in modern times in the field of race matters in the UK.

6.1 The Macpherson Report

The inquiry conducted by the team under Macpherson explored incisively the events that occurred following the murder of a young black man in broad daylight at a bus stop in South London; its research brought together various dimensions of racism—street racism and cultural racism; overt racism and covert racism; direct racism and indirect racism; individual racism and institutional racism. The report chose to attribute the primary cause of police failures to 'institutional racism'—a term first used in the iconic book *Black Power: The Politics of Liberation* by Stokely Carmichael (later called Kwame Ture) and Charles Hamilton (1967, p. 4) (see 'Definitions of racism and race' in Chap. 4). The report provided a definition of 'institutional racism', suited for a particular purpose—inducing

© The Author(s) 2017
S. Fernando, *Institutional Racism in Psychiatry and Clinical Psychology*, Contemporary Black History,
DOI 10.1007/978-3-319-62728-1_6

the police to change their ways of working in order to minimise discriminatory practices and racial bias in the police force:

> The collective failure of an organisation to provide an appropriate and professional service to people because of their colour, culture or ethnic origin. It can be seen or detected in processes, attitudes and behaviour that amount to discrimination through unwitting prejudice, ignorance, thoughtlessness and racist stereotyping that disadvantages minority ethnic people. (Home Department 1999, p. 28)

Macpherson's definition of 'institutional racism' focused on the practical effects of racism—'failure of an organisation [in this case the Metropolitan Police responsible for policing London] to provide an appropriate and professional service' (p. 28), almost ignoring the question of personal responsibility of (in this case) police officers or indeed the person in charge of the police institution. A crucial word in the Macpherson definition of racism was 'unwitting'. This word may have been necessary to absolve individual police officers from direct responsibility for their actions—otherwise the police may not have acted on the report—but it is possible that it caused some confusion (see Garner 2010, pp. 102–116 for discussion of the use of the term 'institutional racism' in different settings).

The Secretary of State for Home Affairs, Jack Straw, speaking for the government of the day (the Labour Party was in office from 1997 until 2010) decreed that racism (as interpreted by Macpherson) should be dealt with by institutional changes in policies and procedures in police practices but that these must be accompanied by efforts to identify and weed out institutional racism in all public and private organisations in the UK—professional bodies, corporations and businesses, educational establishments and so on. Some of the changes that the government instituted in police procedures did help to improve the experience of black residents of London, and even outside London, but were far too limited and, as it turned out, were not sustained. In March 2014, Janet Hills, Chair of the Black Police Association at the London Metropolitan Police ('the Met'), was to say that 'the Met is still institutionally racist … there has been no change, no progression' (Dodd and Evans 2014). Doreen Lawrence (the mother of Stephen, murdered 21 years earlier) gave her view in 2014 that 'laws have changed but I think a lot of police attitudes have not changed much' (*BBC News Online*, February 2014).

A consequence of the Macpherson report of 1999 was that it prompted interest in examining various systems in British institutions, including those involving the 'psy' disciplines. Social work and psychiatric professionals pressed for this to be investigated, but it seems that clinical psychologists did not—the question of institutional racism in clinical psychology is discussed later on in this chapter ('Race matters in professional associations') and in Chap. 8 ('Racism of a psychology report'). The Royal College of Psychiatrists (RCP) appointed an external consultancy to carry out an independent review of institutional racism within the College's structures (see Cox 2001); and two papers (Sashidharan 2001; Cox 2001) in its official organ (the *Psychiatric Bulletin*) attempted to open a debate on institutional racism within British psychiatry. Unfortunately, the expectation that the clinical work in mental health would become less racist institutionally as changes took root (for example at the level of raising awareness of racism and cultural issues among trainees), proved to be unfulfilled. However, it should be stated that many mental health professionals are now aware that professional practice is often experienced as racist by many black and minority ethnic (BME) people in the UK: BME is a label usually meaning black and brown-skinned people, but often stretched to include most racialised groups in the UK (see 'Racialisation' in Chap. 5). As a result, many psychiatrists take more care in the language they use in reports and conversations.

Looked at with hindsight, the way 'institutional racism' was defined by Macpherson—or at least the way it was interpreted—turned out to be too restrictive in that it allowed openly racist individuals off the hook (so long as they covered up their racism effectively) and thence allowed 'institutional practice' to be blamed for personal behaviour that individuals should be accountable for. Doing one's duty, abiding by the rules, should not absolve personal responsibility for racist practices—but this is what sometimes happens once 'institutional racism' is quoted as a reason (or excuse) for failings, in cases when black users (patients) of the mental health services voice complaints alluding to racist practice. Today, the over-representation of black people as schizophrenic continues to be a major problem (see 'Explanations for "schizophrenia" in black people' and 'Racialisation of the schizophrenia diagnosis' in Chap. 5). Sometimes this arises from direct personal prejudice towards black people, in which case, the professional(s) concerned should be held accountable for being racially discriminatory, if this can be done under the existing race relations legislation (currently the Race Relations Act 1976). However much of the injustice of over-representation is attributable to institutional practice, what one might call 'normal practice' in terms of the way diagnoses are made. In such a situation, especially if (as is often the case) the professionals concerned take no interest in trying to minimise the effects of institutional racism or to explain to patients/legal professionals the limitations they may be working under (in carrying out their responsibilities under the Mental Health Act and adherence to 'normal practice'), those professionals (the people making diagnoses or assessments) should be accountable to their professional body. Unfortunately the bodies that are most often involved, the RCP and the British Psychological Society (BPS), are not inclined to confront institutional racism adequately (see 'Race matters in professional associations' later in this chapter).

6.2 Transcultural Psychiatry in the UK

The struggle against racism in mental health services in the UK in the 1980s and 1990s owes much to activities in the field of 'transcultural psychiatry' by members and supporters of the Transcultural Psychiatry Society (TCPS). Most of the critical articles and books on race and culture written during the 1980s were authored by members of this body; and the British voices that were heard, both in the UK and abroad during that period came directly or indirectly from the TCPS. The TCPS has its origins in an informal group of mental health professionals in Bradford (UK) led by Philip Rack, a British (white) psychiatrist, who met together to study issues around culture; Dr. Rack and his colleagues in Bradford thought culture was the main problem behind difficulties that mental health professionals had in relating meaningfully with some of the Pakistani and Polish people living in the catchment area of Lynfield Mount Hospital (Bradford). They stimulated other similar groups, notably in Edinburgh (Scotland) under the aegis of the Professor Morris Carstairs, and finally a national movement took off in the late 1970s to become established as the TCPS, with a membership composed of professionals in the field of mental health, in social work, clinical psychology and psychiatry.

From its beginning, the TCPS included black and white professionals from various cultural backgrounds. The society organised meetings in mental health settings in British cities where immigrants from non-Western ex-colonial countries had settled. The focus of TCPS' discussions was at first 'culture', but that shifted in the early 1980s to issues of 'race' (as well as culture), a move influenced by what patients and clients of the National Health Service (NHS) felt about *their* experiences of what they saw as racism in the mental health system. The discourse among members of the

TCPS and at meetings was influenced by British cultural studies, for example that carried out by the Birmingham group led by Stuart Hall and colleagues—*Policing the Crisis* (Hall et al. 1978) and *The Empire Strikes Back* (Centre for Contemporary Studies 1982). When Philip Rack vacated the chair of the TCPS in 1983, psychiatrist Aggrey Burke (of African-Caribbean origin) took his place in 1984 and shortly after that I became the society's secretary. A new constitution was devised (TCPS 1985) to specify that opposition to racism was the TCPS' primary object. From then, the aims of the society in counteracting racism dominated many of its open meetings, which played a significant role in highlighting the role of 'race' and 'culture' in the mental health field. A short account of the TCPS (which lasted until March 2008) is given in a publication by Diverse Minds (a division of MIND) (Vige 2008).

In the 1980s, the TCPS was deeply involved in lobbying for change in mental health services by addressing both racism and cultural insensitivity in those areas, especially in the practice of psychiatry and clinical psychology. It did so through publications by its members and the content of its national meetings, as well as by lobbying managers of individual NHS service providers. Thus developed a movement that was critical of both psychiatry and psychology, from a race and culture perspective, and which achieved some influence both in the UK and other countries—see Moodley and Ocampo (2014). The transcultural psychiatry that developed in the UK had the makings of a type of psychiatry that was 'critical' of traditional psychiatric practice and was meant to work towards minimising racism in the mental health services. However, this movement was not looked at kindly by the forces at work within the (largely white) professions. Its activities were interpreted as 'troublemaking' and membership of the TCPS was shunned by some BME professionals who were concerned about their careers—especially young psychiatrists and psychologists who strongly desired to climb the difficult ladders in their professions in the hope of breaking through glass ceilings of racial discrimination. At the level of the NHS, too, the type of transcultural psychiatry promoted by the TCPS—essentially an anti-racist approach—was not generally approved of and its influence was undermined.

As pressure developed from minority ethnic groups in some parts of the UK (especially in London (where multiculturalism as a social model was accepted politically) for mental health services to be 'transcultural', some health authorities designated particular posts (for example of consultant psychiatrists in an area with a relatively large black population, such as Tower Hamlets) as nominally 'transcultural'. However, the word 'transcultural' was often interpreted by the authorities in charge of services in a way that dropped the 'race' dimension. More importantly, professionals appointed on the understanding that they would deliver 'transcultural psychiatry' were mainly white doctors who merely voiced an interest in 'culture'; and prospective appointees from black and minority ethnic (BME) backgrounds were excluded as unsuitable on various pretexts. In a particular instance (of an acquaintance of mine who told me his story), the person appointed to a nominally 'transcultural' post was discriminated against to such an extent soon after his appointment that he resigned within weeks. A white psychiatrist was appointed to replace him and the story was put about that the authority was unable to find a suitable person from an ethnic minority.

Gradually many consultants and other senior staff in areas with relatively high number of minority ethnic people began to claim an interest in 'transcultural psychiatry', some of them actually claiming to practice 'transcultural psychiatry', although many had little idea of what it meant. The notion was circulated in the NHS that all psychiatrists should be 'transcultural' and some health authorities claimed that the psychiatrists in their regions were exactly that. By the early 1990s, the term 'transcultural psychiatry' was demeaned in practice as merely meaning a voicing of sensitivity to 'culture', usually (for example) by using interpreters when the clients concerned were not proficient in the English. The suppression of 'transcultural psychiatry', as developed in the UK to work towards minimising racism in the mental health services, combined with the operation of glass ceilings of racial discrimination in NHS appointments during the 1980s, was an indication of institutional racism in practice. The term transcultural psychiatry, as something 'special' that addressed the inequalities that resulted from racism and misunderstandings because of cultural difference, was gradually dropped in centres of excellence [*sic*], as they claimed to 'mainstream' transcultural practice or substitute it with the term 'cultural psychiatry'. The glass ceilings that resulted in a dearth of black and Asian psychiatrists in senior posts at prestigious institutions began to break down in the mid-1990s when (for example) a few black senior psychiatrists were appointed to the IOP (see 'Black people in white-dominated systems' in Chap. 7) and gradually 'transcultural psychiatry' lost its attraction as a means of changing the mental health system in the UK.

6.3 Action on Apartheid

The struggle against racism in the form of improving the experience of black people caught up in the mental health system did not seem, to members of the TCPS, to be disconnected from the struggle in the wider world outside that system. Several TCPS members were active in anti-racist action, including activities to oppose apartheid in South Africa (see 'Discrimination, diagnosis and power' in Chap. 5). In the 1980s, the TCPS collaborated with some psychiatrists sympathetic to the anti-apartheid movement to form an informal group called Psychiatrists Against Apartheid (PAA); the group pressed the RCP to pursue action within the World Psychiatric Association (WPA) to exclude South African psychiatrists (the latter were all-white at the time) from its membership until such time as they supported changes in the organisation of mental health services in South Africa. The result of a longish campaign was that the RCP decided to send a team from the College to report back on the situation in South Africa. This team refused to consult with the exiled members of the African National Congress (ANC) living in the UK that some TCPS members were acquainted with, and refused to look at evidence on race discrimination by South African psychiatrists that TCPS was able to provide. Needless to say, the RCP's alleged fact-finding mission was a stitch-up with their colleagues in South Africa. After being apparently reassured that conditions for black patients in South Africa were satisfactory and having enjoyed tourist trips, the College team advised that no action should be taken by the RCP. It was significant that during one of the discussions at the RCP, an ex-President of the College objected to taking action against 'kith and kin' (the white psychiatrists of South Africa) and went on to allege that PAA members were communist sympathisers—'Reds under the bed' was a popular theme at that time, during the Cold War. In spite of the RCP's taking a white supremacist approach, the TCPS did get somewhere in its campaign. Since there was an academic boycott of South African academia in force at the time, the TCPS lobbied psychiatrists with links to South Africa to keep to the boycott—with some good results.

6.4 Action by Black Professionals

In the 1960s, African-American professionals in the USA came together to devise a strategy known as 'black psychology' that addressed the particular needs of black clients and patients who sought help for mental health problems. According to Watson (1973), who, incidentally, regretted the need for a 'black psychology' but appreciated why it had come about, this movement addressed three areas of concern: first, black psychologists could provide a picture of black family life that is different from that presented by conventional wisdom in the 'psy' disciplines (see 'American social studies' in Chap. 4), emphasising the strengths within that family structure and its ways of making sense of the world that African Americans lived in. Second, in highlighting the excessive numbers of black people being diagnosed as mentally ill, the movement focused on white racism as the cause for black mental illness. Watson believed that 'Blacks have chosen this [approach] because in so doing they have been able to point to white racism itself as a sickness' (p. 16). Third, questioning the validity for black people of established IQ tests, black psychologists devised new tests geared to the black experience. In the field of psychiatry too, black professionals formed an association—the Black Psychiatrists of America separating themselves from the white psychiatric establishment—something that British people have not done; the TCPS, which came nearest to an association to highlight racism in the mental health system of the UK (see above), always included white people in its membership. Attempts in the UK by black and Asian psychologists and psychiatrists ('Asian' refers here to people of South Asian origin) to oppose racism within their professional practices have been few and far between. In practice, any individual who does this becomes marginalised within his or her respective profession and there is little in the way of supportive organisations among these professional groups that they can have recourse to.

Although black professionals in the UK who came together in the TCPS (see above) were concerned about racism in psychiatry and clinical psychology, much of the work emanating from the TCPS was in developing a body of literature that explored those issues in the 'psy' disciplines that led to BME people in the UK often experiencing clinical practice as being racist and culturally dissonant with their backgrounds. Some remedies for these problems, and ways of changing clinical practice, were contained in the literature—the author himself suggested some in a book (Fernando 1988) published nearly 30 years ago—but the TCPS itself did not develop any specific polices to counteract racism, although its members, acting as individuals, participated with others in statutory agencies and professional bodies that did this.

In 1987, the RCP established a committee to consider 'problems of discrimination against trainees, other doctors in psychiatry and patients on the grounds of race' (RCP 1989); but the committee's report was ignored by the governing body of the RCP. Later, in February 2001, in an attempt to get the ball rolling towards cultural and racial sensitivity in the training of psychiatrists, a special group within the RCP, brought together by the then president (who had in the past been a member of the TCPS), worked out a basic structure for training in transcultural psychiatry. But all that happened was that the subsequent President of the College, who had declined to participate in the discussions on changing the training curriculum, despite claiming to be interested in 'cultural psychiatry' (not 'transcultural psychiatry', which had an anti-racist connotation—see above under 'Transcultural psychiatry in the UK') did not take forward the suggestions for that curriculum, which had been agreed on earlier.

All through the 1990s, successive Biennial Reports of the Mental Health Act Commission (MHAC), a government inspectorate that reported directly to Parliament, identified the needs of black and ethnic minorities as a priority, quoting the disadvantages suffered by black people in Britain because of racism (MHAC 1991, 1993, 1995). And the Standing Committee on Race and Culture of the MHAC insisted on statements highlighting institutional racism in the mental health system being included in these reports. The reports likely played a significant part in persuading ministers of the need to counteract racism in mental health services, which led to the funding provided by the Tory government for the Ipamo project (considered later in 'Illustrations of institutional racism'). Also, as chairman of the committee referred to, I was involved with the then Commission for Racial Equality (CRE) to contest psychiatric practice as discriminatory because of the fact that service users face discrimination once they get caught up in the system. This was on the basis of the Race Relations Act (1976), which had made it illegal to discriminate on the grounds of race in providing services. A staff member of the CRE and I met with staff from the Department of Health (DOH) to discuss access to data, but the DOH refused to cooperate. We decided that we were unable to proceed further, but it made the senior staff of the DOH think hard and likely played some part in driving government ministers to look at ways of alleviating the racist nature of psychiatric practice.

6.5 The Black Voluntary Sector (BVS)

In the 1980s and 1990s, the UK became well known in Europe for having a strong so-called 'voluntary sector' (now known as the 'third sector') providing mental health services for BME people. The context in which this came about was as follows: (a) there was government recognition of the statutory sector's failure to provide services that BME people found satisfactory—a fact around which a large literature had developed; and (b) the official approach of government at the time (between 1980 and 2000) was to promote a brand of multiculturalism that promoted the existence, side by side, of different cultural groups enjoying their different ways of life. (The arguments for and against this approach to social policy is beyond the scope of this book.) Since the early 1980s, many counselling services designed for BME people, of South Asian, African-Caribbean, Chinese, Somali, North African and other backgrounds, were evident in many British cities, such as London, Birmingham and Bradford, and were nearly all funded by local authorities and even the NHS (see for example, Fernando 1995, 2005; Fernando and Keating 2009).

I was personally involved with several of these BVS services and remember meeting visitors from Europe who had come specifically to visit them in the hope of replicating them in other parts of Europe. Some of these services followed traditional Western methods of therapy—for instance Nafsiyat followed psychoanalytic methods and had some therapists who spoke languages other than English, and others who claimed to be 'culturally sensitive'. There were various interpretations of what 'cultural sensitivity' consisted of but at Nafsiyat it largely meant largely translating psychoanalytic concepts into various languages and attempting to learn from clients about what these meant in terms of their own culturally determined thinking.

Unlike Nafsiyat, some other projects, such as the Nile Centre and Harambee, provided a range of social and psychologically supportive services with an anti-racist bent (for discussion of these projects see Fernando 2005). However, very few openly advertised themselves as 'anti-racist', often because of a fear that they would lose their funding if they did. Since 9/11, government policies towards funding services meant for specific ethnic or cultural groups has changed—the government now prefers to fund only services that promote 'cultural integration' and are suspicious of services designed for any one ethnic group. The BVS has diminished considerably since about 2005 and is now almost extinct—another instance of institutional racism. The BVS organisations were usually subjected to many more checks than white organisations were subjected to and often made to feel that they were not fully trusted. Some attempted to develop links with university departments but when they did—for instance by providing placements for students—they usually got a bad deal, seldom getting paid for having students (although this happened to white organisations in voluntary sector too) and were often subjected to critical comments implying that they were not up to the standards of white organisations. One such situation, which amounted to colonial-style exploitation is described later in this chapter (see 'Racist exploitation of a black organisation' under 'Illustrations of institutional racism').

6.6 INSTITUTIONAL ACTION

In 1983, a new Mental Health Act came into force in England, Wales and Northern Ireland, whereby an inspectorate, the MHAC (referred to earlier this chapter) was installed. When the lack of commissioners (as members of the MHAC were known) who were from BME communities was highlighted by the TCPS, two black psychiatrists (at the time the term 'black' included people now referred to as 'Asian'), myself included, were appointed to serve as commissioners in 1986. One of the main functions of the commissioners was to visit locations where patients were held compulsorily and report any problems they may find. Soon after being appointed, I discovered that commissioners (who until then had been all-white) had never reported finding any evidence of racism, although many were aware of the discussions in the public domain of possible problems around racist attitudes in the course of 'sectioning' (compulsory detention under a section of the Mental Health Act) and diagnosis. However, my black colleague and I were told by black patients about racist abuse by staff and also of racial discrimination in the way they were dealt with by those staff, including doctors. Generally patients did not want to pursue official complaints because they felt that such action would rebound on them. On discussion within the Commission, it emerged that personal discrimination represented merely the tip of the iceberg of racism in the mental health services. A lobby developed within the MHAC centred on these issues and this became its official 'Standing Committee on Race and Culture'. Successive reports by the MHAC (MHAC 1987, 1991, 1993) highlighted serious problems of racism in the mental health services. Racism in the MHAC itself is discussed later in this chapter in the subsection 'Racism in a government body' under 'Illustrations of institutional racism'.

In 1993 (the same year as that of Stephen Lawrence's murder—see under 'Macpherson report' earlier in this chapter), a report was issued on the inquiry into the death of a young black man called Orville Blackwood and two other African-Caribbeanpatients at Broadmoor (Secure) Hospital (SHSA 1993). The report (the Blackwood Report as it came to be known) identified racism as a major factor in the events leading to all the deaths. While discounting evidence of overt blatant racism, the report described 'a culture within the hospital that is based on white European norms and expectations. As such, there exists a subtle, unconscious on the whole, but nevertheless effective form of organisational racism' and referred to this as 'subliminal racism, where preconceived stereotypes play as great a part as individual needs [on treatment and management]' (p. 54). As one of several specific recommendations, the report recommended 'further research into the diagnosis of schizophrenia in Afro-Caribbeans… [and monitoring of] patterns of diagnosis among minority ethnic groups in the special hospital system' (p. 52). The hospital authorities rejected the report and the government body that commissioned it ignored it, even failing to print sufficient numbers of copies of the report to meet public demand. The chairman of the inquiry team, Professor Herschel Prins, a commissioner who worked with me on a MHAC team, told me that he was told informally that once he left the MHAC, he would be barred from visiting Broadmoor Hospital.

6.7 Government Action

In 2001–2002, the DOH commissioned a new report on what needed to be done about ethnic issues (essentially racism and insensitivity to cultural difference) in the mental health field. To the surprise of many BME people who were involved in campaigning to redress racial inequities in the mental health field, the person asked to produce the report was (and is still) a well-known critic of government policies, the psychiatrist Dr. Sashidharan. After much delay (and it seems much argument with civil servants at the DOH about what should be put into the report), *Inside Outside* (NIMHE 2003) came out, outlining where radical changes were required both in the statutory services (the 'inside') and in work to empower BME communities and the voluntary sector (the 'outside') to voice their concerns and get involved in making changes to those 'inside' services.

The report *Inside Outside* called for policies and action directed at empowering BME communities—the 'outside'—through promoting community development (for discussion of 'community development' see Ledwith 2011). Hope that changes would be made was high among BME people but, even while the author of *Inside Outside* was preparing to put together a team to plan the implementation of the report's recommendation, the DOH approached another party, from a different place and a different background, to devise the implementation plan. To the surprise of many BME people (such as myself and Dr. Sashidharan), an academic who had previously shown little interest in the mental health field produced a report called *Delivering Race Equality* (DOH 2003), also known as 'DRE 2003', drawn up in a university department without proper consultation with BME communities or with any of the professionals who had worked for many years on identifying what kinds of changes were required.

The report known as DRE 2003 was couched in similar terminology to that used in *Inside Outside* but with very different implications. The emphasis on changing statutory services so that they would be in line with what BME communities wanted (and empower them to voice their views) was shifted into an emphasis on collecting information. And, even more importantly, instead of 'community development' there appeared 'community engagement'—the idea being that BME communities needed to engage with services more effectively; the critique of services contained in *Inside Outside* was not pursued. Meanwhile another reported inquiry emerged (the David Bennett Inquiry) and found that institutional racism was to blame for some of the events leading to the violent death of a black patient (David Bennett) while being physically restrained in a mental hospital (Norfolk, Suffolk and Cambridgeshire Strategic Health Authority 2003). A revised version (DOH 2005), or 'DRE 2005', was issued incorporating the government's response to the David Bennett Inquiry—and largely rejecting most of the recommendations in the Bennett Inquiry Report itself (which had blamed institutional racism for much of the mishaps that led to David's death). The author of the report *Delivering Race Equality* was made a member of the House of Lords. Soon afterwards he resigned his leadership of DRE. The DRE process itself hit trouble from its early days (Fernando 2006) and failed to achieve any of its objectives. Essentially it was a waste of money. The whole DRE initiative, from the way it was instituted to the way it was managed, strongly suggested institutional racism. Some of the problems of DRE have been described in detail elsewhere (Fernando 2009; RawOrg 2011).

Once the poor outcome of DRE sank in, BME people seemed to lose enthusiasm for the struggle to bring about change, losing confidence in the honesty of the powers that be and disillusioned with the individual black and brown-skinned people chosen by them as 'leading' change—two of whom had been elevated to membership of the House of Lords. Receiving honours seemed to be associated with not voicing much criticism of government policy and keeping silent on issues of racism; yet another aspect of institutional racism. At a government level the view had apparently developed that nothing more should be or could be done, but even more significantly, that 'race' issues were in the past—the notion of Britain being a post-race society (see 'Obama years' in Chap. 8).

6.8 Illustrations of Institutional Racism

I describe here four illustrations of institutional racism in the mental health field: two projects which were set up to be staffed mainly by black people in the hope of minimising the everyday racism that users of mental health services may experience; a voluntary organisation (in the BVS) that developed an association with a university department; and aspects of the struggle within a statutory organisation set up by the MHAC. These accounts all refer to events in the 1990s.

6.8.1 The MOST Project

The results of a survey carried out at the Maudsley Hospital in the 1980s (Moodley 1995) showed that: 'While white patients considered the social contacts made through hospital as being more important, African and African-Caribbean patients rated seeing a member of staff of their own colour, being understood and receiving help with finding jobs as more important' (p. 128). At the time there was concern about many African-Caribbean patients not complying with the medication prescribed by their (mostly white) doctors at the IOP/Maudsley Hospital. Special funding was obtained by the IOP for a community-based service to which patients who were 'non-compliant' (with medication therapies) could be referred by the doctors at the Maudsley. The Maudsley Outreach Service Team (MOST) was set up in the early 1980s with Dr. Moodley, a psychiatrist of black South African background, as the director. In the only reference to the work of MOST in published literature (see next paragraph for why that is so), Dr. Moodley (1995) later wrote: 'We believe that our success was a result of working *with*, rather than for or at, our patients. Our interventions were always made explicit and nearly always agreed upon between the professionals and service users involved—with compromises on both sides' (p. 138, emphasis in original). I know that many of the African-Caribbean people who were fortunate enough to obtain the services of MOST appreciated the help they got from the service. But, when the service came up for consideration, the Maudsley decided to change its ways of working and did not stabilise Dr. Moodley's appointment as its consultant. In 1990, the work of MOST was 'mainstreamed' by the Maudsley Hospital, which meant that its functions were placed within the main mental health service structure of the hospital. Once this happened, the style of working at MOST changed although MOST had been established precisely because the system of mental health care administered by the Maudsley had been experienced by black people as unsatisfactory. But the transfer of management resulted in another serious consequence.

Dr. Moodley and her staff had accumulated data on the work of MOST, hoping to analyse the data and publish the findings in due course. Once the service was placed under the direct management of the Maudsley, the IOP (which was associated with the hospital) demanded all the records and research data on the past activities of MOST, claiming these to be its property, and intimating that its researchers would analyse and publish the outcome data. But the analyses (if done) were never published, so there is no published record available of the work of MOST. Dr. Moodley and her staff considered that if the data had been analysed, the outcome data would have shown how unique their service had been, at least in terms of the satisfaction of the (mainly black) service users; but perhaps even more so in terms of showing better clinical outcomes than the Maudsley's other services. Valuable research was therefore thwarted by the IOP, most likely because it may have reflected poorly on traditional practices at the Maudsley, but also because of institutional racism—MOST was seen as a project that served mainly black clients and many of its staff, and notably its lead consultant, were black.

In 1998, I had a call from a senior medical academic at the IOP who was advising the DOH on a report called the National Service Framework for Mental Health (DOH 1999) asking whether I knew of any evidence available on ethnic minority projects that had shown good outcomes. I mentioned the MOST project and told him that the data on its work was taken by the IOP but never published. The caller did not consider that my views and a book-chapter by Dr. Moodley (1995) was sufficient 'evidence' to be quoted in a government report. He was also a colleague of the professor at the IOP who had demanded the data on the MOST project.

The impression still persists that there have been no mental health projects for BME people that have shown good outcomes, and I wonder how many others there have actually been that have never achieved 'evidence' status. Institutional racism is not just unjust but bad for research and progress.

6.8.2 The Ipamo Project

In the early 1990s a project was developed with special funding from the DOH to provide a service for black people in the Brixton area of South London. The aim was to set up facilities for inpatients independent of the statutory service, and staffed entirely by black people, in line with ways of working that were deemed to suit their needs and the particular problems they faced (which the statutory sector was failing to address). In mid-1995, a voluntary organisation (non-governmental organisation, NGO) was established, called Ipamo, which was to work in partnership with the statutory bodies that provided the local mental health services for Lambeth (which included Brixton) in South London. The arrangement was that the statutory bodies held responsibility for the funds and capital development while Ipamo, through its Board of Directors, was responsible for developing and running the service. Being a member of Ipamo's board, I was privy to all the deliberations that took place over the next two years.

The ethos of the prospective service (for black patients) to be provided by Ipamo was worked out in the meetings and conferences it sponsored. The Board of Directors approved the working practices which were to be implemented by Ipamo's director, Malcolm Phillips, an accomplished mental health professional and psychologist who was well respected in black communities. The vision for Ipamo was a model of care that brought in traditional African approaches to mental health; ideas from black social and political movements, drawing on work from the USA; and notions of spirituality pursued by local black churches. A house was bought and architects appointed redesign and refurbish it in a style suited to British black communities. What I saw happening as we worked on the Ipamo project was worrying. It seemed evident to me that our plans were being undermined by psychiatrists attached to the statutory services and the managers of those services who held control of the funds. It was not done openly, but through actions such as picking quarrels with the architects who were designing the prospective building for the service, even threatening to take them to court; by raising objections to the way the director of Ipamo (supported by the Board of Directors) intended to manage the service, with no white psychiatrists involved; and so on. Ultimately, the attempted collaboration between the statutory sector and Ipamo broke down and the statutory authority called off the project, deciding to use its special funding elsewhere. What had happened, by means of institutional processes, seemed to be that the white establishment of the statutory service had successfully prevented a service for black people from developing—a reflection of institutional racism. We heard later that the money originally designated for Ipamo was redirected to support a psychiatric forensic service—ironically, this was the sort of service that black people tended to feel was most racist.

6.8.3 *Racist Exploitation of a Black Organisation*

An organisation in the BVS providing psychoanalytic psychotherapy for BME people was founded in 1983 by a good friend of mine at the time, Jafar Kareem, a British-trained therapist. In early 1992 Jafar died very suddenly. A new director was appointed after considerable delay; and several years later I was invited by the Board of Trustees of the project to get involved in its management—something I agreed to do, hoping to widen its scope in the way I had already discussed with Jafar before his death. Within a few months of getting involved, I became the chairman of the Board of Trustees of the organisation, and I then realised that there were serious administrative problems that the trustees had to contend with. Of the many issues that required attention, I have chosen to describe one that illustrates an aspect of institutional racism that many organisations in the BVS suffered from in the 1980s.

The organisation concerned was funded by the local authority of a London borough together with a health organisation (part of the NHS) to provide psychological therapies for people identified as belonging to BME communities. Over the previous few years, it had become associated with a university department in running a course (approved by the university) for an MSc degree. This course was advertised as being jointly organised between a department of the university and the voluntary organisation. What had actually happened was that the (black) director of the voluntary organisation (without any consultation with its Board of Trustees) had agreed an arrangement with a professor of the university to provide teaching and supervision to university students registered for an MSc. The director had been given an honorary post by the university and some facilities and staff time (of the voluntary body) were devoted to the university course. When, as chairman of the Board of Trustees of the voluntary organisation, I asked for details of the arrangement between the two organisations, I was told that it was a gentleman's agreement between the director of the voluntary organisation and the professor attached to the university who was also a local NHS consultant psychiatrist. The crunch was that all the fees paid by the students—and these were not inconsiderable—were pocketed by the university; nothing came to the voluntary organisation, although the students used its facilities (including even photocopying facilities) and were being tutored by some of its therapists, who were paid by the voluntary organisation. When called on to explain, the university authorities took the line that the black voluntary organisation obtained hidden benefits from being associated with a prestigious university. The white institution (the university concerned) got the money and the black voluntary organisation did (much of) the work. And the black clients were short-changed. Eventually, the joint course (at that time the only one of its kind in the UK) was discontinued because the university authorities were unwilling to change the colonial arrangement that had been instituted. Colonialism and institutional racism coincide in many ways.

6.8.4   Racism in a Government Body

I referred earlier to the MHAC, of which I was a member between 1986 and 1995. The MHAC was organised as 'visiting teams'—groups of commissioners that visited locations where people were compulsorily detained. I chaired for a while the Standing Committee on Race and Culture within the MHAC. The problems that surfaced within that body illustrate the nature of the struggle that often takes place (usually behind closed doors) in many organisations.

I joined the MHAC in 1986. In 1989, a new chairman was appointed to head the organisation. The 'working party on black and ethnic minorities', which I chaired at the time, was renamed the 'National Standing Committee (NSC) on Race and Culture', as all working parties were renamed NSCs. In October 1989, all commissioners received a memorandum from the commission chairman stating that in future all NSCs would be chaired by persons appointed by him and these chairpersons would form the Central Policy Committee (CPC) of the Commission that would determine MHAC policy. When the names of the selected chairpersons were announced a few days later they turned out to be the people already in post (having been elected by member of their respective NSCs) except in the case of the NSC on Race and Culture, which was given to a new chairperson—a white man who had previously not been involved in the committee at all. The result was that the CPC (formed by the chairpersons of NSCs) would be all-white.

Suspecting unfair practice, I telephoned the commission chairman who told me that he wanted a white person to chair the Race and Culture NSC because (as he put it) people listen to white people more than to black people. Having made a note of the telephone conversation, I then undertook a struggle which involved (a) obtaining legal advice on the applicability of the Race Relations Act (1976) to the action of the commission chairman and making this known to members of the Central Policy Committee (CPC), the governing body of the MHAC at that time; (b) lobbying individual members of the CPC and apprising them of my treatment by the commission chairman; and (c) making the issues (as I saw them) known unofficially to persons at the DOH, knowing that this information would become part of the general 'gossip' there. At the end of this, the commission chairman changed his earlier decision and nominated me as the chairman of the NSC on Race and Culture; rescinded his earlier decision to form a new style CPC; and wrote to me stating that he has been instructed by the CPC to apologise to me unreservedly. I replied stating that this was not enough, implying that I was considering further action. A few days later I had a telephone call from a prominent member of the MHAC well known to me.

The person who telephoned me put it to me that although the MHAC chairman had taken what seemed to be a racist action, he was 'not really' racist but just 'paternalistic' (see reference to statement by Easterly (2006) in section 'Transformations after WWII' in Chap. 4). Also she sounded me out on what might satisfy me—presumably in order to prevent details of this racially charged event in the MHAC getting into the public domain. I suggested an official policy for the MHAC that would render future racist actions unsustainable and ensure monitoring of staffing arrangements within the MHAC from an anti-racist viewpoint. The policy—'Race Policy'—was agreed by the Commission and published in the fourth and fifth commission reports (Mental Health Act Commission, 1991, 1993). The Race Policy itself was revised just before I left the Commission in 1995 and was renamed 'The Policy on Race and Culture' (Mental Health Act Commission, 1995, pp. 210–212).

The Race Policy helped when the NSC on Race and Culture, in the course of monitoring that policy, demanded a proper ethnic mix of membership of the CPC—and got it. But the question at the time was about sustainability of the change towards anti-racism. The seventh commission report (Mental Health Act Commission 1997), issued after I left the MHAC, did not refer to a policy on race at all but merely to the Commission having identified 'three target areas' as a 'focus during visits', namely ethnic monitoring, interpreters and racial harassment (p. 158). The anti-racist approach epitomised by the Race Policy that was operational from 1991 until 1995 has since been abandoned in favour of a 'culture-sensitivity' and 'equal opportunity' approach. Gone, it seems, was the need to keep the Commission 'free of discriminatory practices on racial grounds' (MHAC 1995, p. 212) and to monitor the actions of the Commission itself for institutional racism. A wider consideration of my experiences as a member of the MHAC, including more details of the issues given above, are described elsewhere (Fernando 1996). The organisation of the work of the MHAC, later subsumed within the Care Quality Commission (CQC) has changed considerably since 1995—monitoring racism no longer seems to be within its remit and the mental health work of the CQC is now seen as relatively weak with respect to highlighting issues of a fundamental nature. The minor institutional change I managed to bring about in the MHAC was not sustained. Institutional racism triumphed after all. And, as I have observed in many other occasions, in this instance too, institutional racism was fronted by black or brown-skinned people.

6.9 Race Matters in Professional Associations

The professional bodies in the UK that cover clinical practice in psychology and psychiatry are the Division of Clinical Psychology (DCP) of the BPS (referred to earlier) and the RCP (also referred to earlier). I have little experience of the BPS and DCP, but since I practised as a psychiatrist for over 20 years I belong to the RCP. None of these bodies has shown much interest in addressing racism in clinical practice and in the underlying theory relating to mental health, but the institutional racism that appears to be present in them are illustrated by the following stories.

I came into contact with the DCP in connection with the aftermath of the report issued by this body, described later in this book (see section

'Racism of a psychology report' in Chap. 8). I was told then that a 'Faculty of Race and Culture' had been in existence (as part of the DCP) for several years but had been disbanded a short time before the report *Understanding Psychosis and Schizophrenia* was issued. It seems that if this faculty had been active at the time, the report would have been examined by it before its publication and the near-racist language and the other faults in the report may have been corrected before publication. Individual members of the faculty told me that the section had been disbanded without any consultation with its members; and that when some of the members met with the chairman of the DCP, he gave them no proper explanation as to why their faculty had been disbanded.

As mentioned earlier (under the section 'Transcultural psychiatry in the UK'), the main body of psychiatrists and clinical psychologists in the UK who were critical of mental health practice during the 1980s and 1990s were in the TCPS (referred to earlier). TCPS members were often very critical of much of the work of the RCP, especially the latter's resistance to addressing racist practice in mental health services—something that several individual clinical psychologist had also been critical of. In the mid-1990s, some of the prominent members of the TCPS (including myself) decided to accede to an invitation of the RCP to form an interest group within it as a step towards a formal 'section' (equivalent to a 'faculty' in the BPS structures) of the RCP. At a well-attended inaugural meeting of this group, the Transcultural-Psychiatry Special Interest Group (TSIG), the main topic discussed by attendees (all of whom were psychiatrists) was the issue of racial discrimination in the RCP itself and in consultant appointments in the NHS. Within a few days of the meeting, the (then recently appointed) chairperson of TSIG received a letter from the President of the RCP stating that the TSIG should not discuss political issues such as discrimination but restrict itself to discussing academic matters. After that initiation to working in the RCP, the TSIG never considered contentious issues around race, confining itself to matters around 'culture'. The TSIG pursed a very low-profile position within the RCP and at the time of writing (mid-2017) has been inactive for at least one year. Discussions of 'race' seem to be off the agendas of both the BPS and the RCP. Institutional racism may well be in the ascendency.

References

BBC News on line. (2014). Doreen Lawrence says sections of police "still racist", *BBC news*-24 February 2014. Retrieved on August 8 from: http://www.bbc.co.uk/news/uk-26321708.

Carmichael, S., & Hamilton, C. V. (1967). *Black power. The politics of liberation in America*. New York: Random House.

Centre for Contemporary Studies. (1982). *The empire strikes back. Race and racism in the 70s Britain*. London: Hutchinson.

Cox, J. (2001). Commentary: Institutional racism in British psychiatry. *Psychiatric Bulletin, 25*, 248–249.

Dodd, V. and Evans, R. (2014). Lawrence revelations: Admit institutinal racism, Met chief told *The Guardian (website)* 7 March, 2014. Retrieved on August 8 from https://www.theguardian.com/uk-news/2014/mar/07/lawrence-revelations-institutional-racism-met-police.

DOH (Department of Health). (1999). *National service framework for mental health. Modern standards and service models*. London: Department of Health. Retrieved on September 28, 2016 from https://www.gov.uk/government/publications/quality-standards-for-mental-health-services.

DOH (Department of Health). (2003). *Delivering race equality: A framework for action*. London: DOH.

DOH (Department of Health). (2005). *Delivering race equality in mental health care an action plan for reform inside and outside services and the government's response to the independent inquiry into the death of David Bennett*. London: DOH.

Easterly, W. (2006). *The white man's burden. Why the West's efforts to aid the east have done so much ill and so little good*. Oxford and New York: Oxford University Press.

Fernando, S. (1988). *Race and Culture in Psychiatry*. London: Croom Helm. Reprinted as paperback Routledge, London 1989.

Fernando, S. (1995). Social realities and mental health. In S. Fernando (Ed.), *Mental health in a multi-ethnic society* (pp. 11–35). London: Routledge.

Fernando, S. (1996). Black people working in white institutions: Lessons from personal experience. *Human Systems: The Journal of Systemic Consultation and Management, 7*(2–3), 143–154.

Fernando, S. (2005). Multicultural mental health services: Projects for minority ethnic communities in England. *Transcultural Psychiatry, 42*(3), 420–436.

Fernando, S. (2006). Blowing in the wind. *Openmind, 137*, 24–25.

Fernando, S. (2009). Inequalities and the politics of 'race' in mental health, In S. Fernando & F. Keating. (Eds.), *Mental health in a multi-ethnic society a multidisciplinary handbook* (2nd ed.). London and New York: Routledge.

Garner, S. (2010). *Racisms: An introduction*. London, Los Angeles and New Delhi: Sage.

Hall, S., Critcher, C., Jefferson, T., Clarke, J., & Roberts, B. (1978). *Policing the crisis. Mugging, the state, and law and order.* Basingstoke: Macmillan.

Home Department. (1999). *The Stephen Lawrence Inquiry. Report of an Inquiry by Sir William Macpherson of Cluny.* Cm 4262-I. London: The Stationery Office. Retrieved on October 10, 2016 from https://www.gov.uk/government/publications/the-stephen-lawrence-inquiry.

Ledwith, M. (2011). *Community development: A critical approach* (2nd ed.). Chicago and Bristol: Policy Press.

MHAC (Mental Health Act Commission). (1987). *Second Biennial Report 1985–87.* London: HMSO.

MHAC (Mental Health Act Commission). (1991). *Fourth biennial report 1989–1991.* London: HMSO.

MHAC (Mental Health Act Commission). (1993). *Fifth biennial report 1991–1993.* London: HMSO.

MHAC (Mental Health Act Commission). (1995). *Sixth biennial report 1993–1995.* London: HMSO.

MHAC (Mental Health Act Commission). (1997). *Seventh biennial report* 1995–1997. London: HMSO.

Moodley, P. (1995). Reaching out. In S. Fernando (Ed.), *Mental health in a multi-ethnic society* (pp. 120–138). London and New York: Routledge.

Moodley, R., & Ocampo, M. (Eds.). (2014). *Critical psychiatry and mental health. Exploring the work of Suman Fernando in clinical practice.* London and New York: Routledge.

NIMHE (National Institute for Mental Health in England). (2003). *Inside outside. Improving mental health services for black and minority ethnic communities in England.* London: Department of Health. Retrieved on October 10, 2016 from http://webarchive.nationalarchives.gov.uk/+/www.dh.gov.uk/en/Publicationsandstatistics/Publications/PublicationsPolicyAndGuidance/DH_4084558.

Norfolk, Suffolk and Cambridgeshire Strategic Health Authority. (2003). *Independent Inquiry into the death of David Bennett* (Chairman: Sir John Blofeld), Cambridge, England: Norfolk, Suffolk and Cambridgeshire Strategic Health Authority NIMHE (National Institute for Mental Health in England). (2003). *Inside Outside. Improving Mental Health Services for Black and Minority Ethnic Communities in England.* London: Department of Health.

Race Relations Act. (1976). London: Her Majesty's Stationery Office.

RawOrg. (2011). *The end of delivering race equality? Perspectives of frontline workers and service-users from racialised groups.* London: MIND and The Afiya Trust.

RCP (Royal College of Psychiatrists). (1989). *Report to council of the special committee on psychiatric practice and training in a British multi-ethnic society.* London: RCP.

Sashidharan, S. P. (2001). Institutional racism in British psychiatry. *Psychiatric Bulletin, 25,* 244–247.

SHSA (Special Hospitals Service Authority). (1993). *Report of the committee of inquiry into the death in Broadmoor Hospital of orville blackwood and a review of the deaths of two other Afro-Caribbean patients: 'big, black and dangerous?' (Chairman Professor H. Prins)*. London: SHSA.

TCPS (Transcultural Psychiatry Society). (1985). *The constitution of the TCPS (UK)*. London: TCPS.

Vige, M. (Ed.). (2008). *Goodbye TCPS*. London: Diverse Minds.

Watson, P. (1973). Race and intelligence through the looking glass. In P. Watson (Ed.), *Psychology and race* (pp. 360–376). Harmondsworth: Penguin.

CHAPTER 7

Persistence of Racism Through White Power

In 1965, the UN (1965) adopted a convention that obliged member states to work towards the elimination of racial discrimination (see 'Transformations after WWII' in Chap. 4); and in the mid-1970s, the UK installed a body called the Commission for Racial Equality (CRE) to implement polices to counteract racism and help individuals faced with racial discrimination. In 2006, the responsibilities of the CRE (established under the Race Relations Act 1976) were taken over by a (new) Equality and Human Rights Commission (EHRC) established under the Equality Act 2006, superseded later by the Equality Act 2010 (TSO 2010). Apart from addressing several other forms of discrimination, the changes in the late 1990s and the years after 2000 represented a toning-down of action to uncover and deal with racism as such, especially since the funding for the organisations meant to help in addressing discrimination was progressively reduced. For instance, the CRE was legally bound to focus on combating racism while the EHRC, established under the Equality Act, was expected to promote equality in relation to 10 characteristics, namely age, disability, gender reassignment, marriage and civil partnership, pregnancy and maternity, race, religion or belief, sex and sexual orientation—in other words, racism was only one of ten types of discrimination that the EHRC is called on to prevent.

In its submission to the UN Committee concerned with monitoring its International Convention on the Elimination of All Forms of Racial Discrimination (ICERD), fifty years after its adoption, the (British) EHRC (2016) reported little that was alarming—all that seemed to be wrong, according to the Commission, was that there were some continuing ethnic inequalities in access to health facilities among people classified as 'Gypsies and Travellers'; and that there was some evidence of excessively high rates of black ethnic groups having a poor experience of mental health services, with respect to length of stay in hospital, and likelihood of compulsory detention. This was a travesty of what was happening on the ground in the field of mental health. A study carried out jointly by the EHRC and the Runnymede Trust (2015), a non-governmental organisation, was damning in pointing to many areas of inequality where the UK has failed to meet ICERD's expectations. Clearly the expectations of the UN have also not been realised more generally—in fact the gains made in Europe and North America in the 1980s seemed to have reversed in the new millennium from about the time of 9/11. As a prelude to discussing the changes since 9/11, I shall consider in this chapter some of the ways in which racism continued during the years between 1965 and 2001 (9/11) although often obscured by people being careful not to use language that may cause offence by appearing to imply racist attitudes—a tendency that has been decried and sometimes condemned, more recently, as 'politically correct language' (see Weigel 2016).

7.1 Controlling Racialised Minorities

It was suggested earlier (see 'Discrimination, diagnosis and power' in Chap. 5) that the diagnosis of schizophrenia is a means by which certain ethnic groups are subject to social control—control here being tied up with disempowering communities of black people by transmitting an image of their being disorganised or unreliable. There are other methods of control too. Four tables (Tables 7.1, 7.2, 7.3 and 7.4) show ethnic differences in situations that affect racialised groups—in this case black people in the UK. These involve the psychiatric system, the policing system, the prison system and the education system; and in all, there are very similar levels of ethnic disproportionality. These are bland statistics and do not explore what lies behind the differences, but all seem to point in one direction—racial inequality.

The information in Table 7.1 is based on age-standardised rates of diagnosis per 100,000 per year made in 1997–1999 in London, Nottingham and Bristol (Fearon et al. 2006). As pointed out earlier, diagnostic labelling carries a high a degree of subjective judgement often based on traditional stereotypes (see 'Racist IQ movement' and 'Alleged mentality of black

Table 7.1 Ethnicity and diagnosis

Ethnic group	All psychoses	Narrow schizophrenia
White British	20.2	7.2
African Caribbean	140.8	70.7
Black African	80.6	40.3

Table 7.2 Ethnicity and 'stop and search'

Ethnic group	'Stop and search' ratios
White:Black	1:21
White:Asian	1:2.5

Source Stop Watch (2017)

Table 7.3 British citizens in prison in England and Wales

Ethnic groups	% in prison	% in general population
White	73.8	88.3
Black/Black British	13.2	2.8
Asian/Asian British	7.9	5.8

Source Berman and Dar (2013)

Table 7.4 School exclusions

Ethnic group	Exclusion from school	
	Fixed period	Permanent
White British	5.14	0.08
Black Caribbean	10.84	0.34
Black African	5.37	0.11
Black other	8.27	0.22

Source DOE (Department of Education) (2012)

people' in Chap. 4; 'Racialisation of the schizophrenia diagnosis' in Chap. 5; and 'Institutional action' in Chap. 6). Table 7.2 was compiled from statistics kept by the London Metropolitan Police for the year April 2015 to March

2016, presented on the website of an independent research organisation called Stop Watch (2017). 'Stop and search' is the name given to a particular activity (obvious from the title) that police officers are empowered to carry out under Section 60 of the Criminal Justice and Public Order Act (1994) (National Archives 1994)—described in *The Guardian* newspaper as 'a most draconian stop-and-search law that plays to police prejudice' (Taylor 2014); and something they are allowed to do without necessarily having to specify why they stop and search. Black people were stopped 21 times more often than white people in the above time period, while people identifying as 'Asian' were stopped 2.5 times more often than white people (the term 'Asian' refers here to people of South Asian origin). The ethnic breakdown of the total prison population in England and Wales, shown in Table 7.3, refers to the year 2013. While merely 2.8% of the general population were black, 13.2% of prisoners were black people, in marked contrast to the rates for white people. Table 7.4 refers to statistics applicable to England in 2009–2010. Black Caribbean children were excluded from school for fixed periods twice as often and were over four times as likely to suffer permanent exclusion. The striking picture from the four tables is that black people, especially those of Caribbean origin, are highly disadvantaged throughout life, suffering control and disempowerment in various ways from childhood onwards. What happens in the mental health system is just one part of a total scene of racism in British systems, whether in mental health, criminal justice or education.

7.2 Employment in the Mental Health System

In May 2016, *The Guardian* newspaper reported that the National Health Service (NHS) 'has moved backwards in relation to race equality' (Jolliff 2016), partly based on a study two years earlier (Kline 2014) that talked of 'the "snowy white peaks" of the NHS', referring to the fact that its senior posts are nearly all held by white people. This lack of progress in opening the workforce of the NHS to ethnic diversity likely reflects a worsening of institutional racism in appointments to well-paid jobs generally. The back story is significant. In 2010–2012, the then Tory-dominated coalition government reorganised the NHS. I was told that many black and minority ethnic (BME) people lost their positions in the course of reordering membership of various committees. However, this was on top of subtle, and at times not so subtle, racism in the employment procedures used by the NHS and most British institutions in both the public and private sectors. A personal story here might illustrate of how racism works out in employment situations at a relatively high level of the NHS.

In the mid-1970s, I applied for a consultant job at the prestigious London (later called the Royal London) Hospital. I had been a senior registrar (generally considered a pre-consultant grade) at the same hospital some years earlier, was well qualified (with a MD research degree for which I had been supervised by the then professor of psychiatry at the London); and I know I was well thought of by colleagues at the London Hospital. At initial visits to meet other consultants, I was given to understand that I was the favoured candidate for the appointment. I was very surprised to be informed after the interview that I was not successful. Within days, four of the senior academics who had been on the appointing panel contacted me personally to say that they thought I should have been appointed—but none of them were willing to tell me (on the grounds of confidentiality) what had happened at the discussion following the interviews of candidates, although two persons hinted that it was to do with the chairman of the panel who was also the chairman of the relevant health authority. I then recalled that the chairman of the interviewing panel had made a caustic near-racist remark about Jewish people when questioning me about my research work which had been about depression among Jewish people. Several years later (1985), I head of what may well have been a similar situation at an appointments committee at Cambridge that had been taken to an industrial tribunal (and hence reported in the public domain).

In 1958, an industrial tribunal decided that an Asian colleague and friend, Dr. Sashidharan, was discriminated against unlawfully at an appointments committee held by the Cambridge Health Authority (CHA). According to the official record noted in a publication issued by the CRE (Anwar and Ali 1987), '[t]he Industrial Tribunal found that Dr. Olive [a panellist on the appointments panel convened by CHA] had said that "it was not possible to be unprejudiced against the Indian doctor" (Dr. Sashidharan); that "it would not be suitable to appoint him because there was no significant [South-]Asian population for him to serve", and that "evident cultural differences made him unsuitable for the post"' (p. 86). Dr. Sashidharan (as in my case) was a brown-skinned man who was well qualified for the post he applied for and was competing with white candidates for a prestigious post in the NHS linked to Cambridge University. An important difference from my own experience at the interviewing panel was that a consultant psychiatrist, Dr. Jean Werner, who was one of the panellists on the appointment committee, told Dr. Sashidharan what had happened at the appointments committee discussions and also gave evidence at the tribunal hearing. In my case, all four panellists hinted at racism at the appointments committee when speaking to me, but none of them were explicit about what had happened and they certainly would not have given evidence at an industrial tribunal. Dr. Sashidharan was fortunate in that a (white) colleague was willing to expose the racism that took place at the appointments committee. The outcome was that the health authority concerned with the associated university ruled the appointments committees out of order but appointed the same people on a new appointments committee. The new committee made the same decision as the previous one had made—Dr. Sashidharan was not appointed. He had great difficulty in obtaining a suitable post for quite some time after that—maybe the news got around that he was a troublemaker.

7.3 Institutional Racism in the Department of Health (DOH)

When the Labour government took office in 1997, the Prime Minister appointed Paul Boateng, a black man, as junior minister in the Department of Health to be in charge of mental health—an appointment that raised the hopes of many black people in the mental health field, aware that this minister had spoken out against racism for many years. This junior minister soon consulted many people from BME communities, including myself, on the nature of the problems faced by black people in the mental health field. However I realised whilst talking with him that he already had a fair idea of the issues affecting black people caught up in the mental health system and the questions these raised around the practice of psychology and psychiatry. In the early days after his appointment, the minister concerned voiced his intention to deal with racist practices and related stories of racism that he had experienced personally in the mental health system. But as the months went by his public statements seemed much less critical of mental health services.

In 1998, the Secretary of State (Minister) for Health, Frank Dobson, announced a consultation for a framework for mental health services, and an 'external reference group' (ERG) to advise on it. The ERG turned out to have no black professionals in it—and many people in general, i.e. service users, had been critical of the services on the grounds or its institutional racism. However two black service users, both well known to me, were appointed to the ERG. They persevered in attending meetings but found that their views were not being listened to and were not represented in the final reports of the ERG. Also, they found the experience very stressful, finding it difficult to have their voices heard and sometimes being insulted. For example, one of the service users told me that a member sitting next to her at one of the meetings told her that she should stay quiet because it was an honour for her to be there among professionals (who were all white). The objections to the exclusion of effective BME voices from among several BME professionals available came into the public domain with an open letter signed by 16 mental health professionals to the Secretary of State for Health—reprinted in *Openmind* (Fernando 1998)— that included the following paragraph:

> This [exclusion of effective BME voices] has been done by a combination of (a) including in the groups many people of high status from prestigious bodies concerned with maintaining the status quo; (b) appointing very few Black and Asian people, mostly without experience of committee work, and some clearly token appointments of people not in a position to voice race and culture issues; and (c) not appointing people who have been in the forefront of advising on and developing strategies to counteract inequalities arising from these issues (some with an international eminence in these areas). We are surprised that a Labour Government which has stated a public commitment to counteract social exclusion should preside over a policy by one of its departments of practising exclusion[.] (Fernando 1998, p. 15)

Later in 1998, the Secretary of State for Health announced consultations on a major review of the Mental Health Act of 1983. A group of BME workers (which included me) got together to discuss possible suggestions, thinking that the revisions would include ways of alleviating problems of racism. To our consternation this was far from what the government proposed. The proposals seemed designed to increase problems for BME people by (for example) introducing community treatment orders and make it easier for 'sectioning' (compulsory detention) to take place through changes in legal wording including the expansion of the (legal) definition of mental illness. All professional groups connected with mental health, except the British Psychological Society (BPS) but including service user groups, objected to these changes; and there was a long period of contentious discussions between civil servants of the DOH and with various groups of professionals and service users—including a BME group that called itself the National BME Mental Health Network (BMENW) composed of both professionals and service users (for some details of its work see Inyama 2009). But finally, a Mental Health Bill 2006 (to amend the 1983 Act) that included nearly all the government's suggestions was put through parliament and implemented in 2007.

In the early part of these discussions (while he still worked at the DOH —he was later moved to the Treasury), the black junior minister I mentioned earlier was prominent in supporting and fronting the government suggestions for changes to the Mental Health Act, but he soon moved to the background and then left the DOH. During these discussions I noticed worrying behaviour among the DOH civil servants indicative of lack of respect for their minister. An Asian civil servant, who had expressed some sympathy for BME problems and advised me on how the BMENW could best raise their concerns in discussions, was suddenly moved to another job. Finally I was told that one of the members of BMENW who worked for a voluntary agency had been given to understand that he might be eligible for a post at the DOH if he toned down his opposition to the changes suggested by the government. To his credit, he did not. In my dealings with the civil servants at the DOH, during discussions about revision of the Mental Health Act between 1999 and 2003 (see above), I found them to be very resistant to acknowledging any possibility that racism plays a part in the problems that BME people have in the mental health services. Apart from those described above, other incidents too have indicated endemic racism in the DOH. When, the report *Inside Outside* (NIMHE 2003) was being prepared (see 'Government action' in Chap. 6) its author found there was resistance by civil servants to including any references to racism in the document (Sashidharan 2006). The changes suggested in *Inside Outside* were not pursued by the government; instead a new report was hastily written, by someone with little knowledge of the mental health field, called *Delivering Race Equality* (see 'Government action' in Chap. 6). All this was indicative (to me) of institutional racism at the DOH.

Individuals who identify as BME and acknowledge their heritage *and* voice anti-racist sentiments publicly are few and far between—many do so in private only. Those who achieve positions of prestige (for example by being nominated to the House of Lords) can wield some influence but many seem to avoid taking an anti-racist stance once they get into such positions, apparently retreating into psychologically gated mindsets, unable (it seems) to provide leadership against racism. What seems likely is that there are powerful socio-political forces in society that select out for 'success' those BME people who generally support the establishment and unlikely to 'rock the boat'—generally regarded as 'token-blacks'. My experience in the politics of the mental health field is that when an official body is set up to consider issues of race and culture, it is either these 'token blacks' who are invited to participate or people (a) carefully chosen for being 'safe' from the point of view of the establishment, which generally wants to suppress any views that may challenge powerful professional groups or government policy; and (b) who seem to follow personal rather than community agendas. In other words, the way institutional racism operates has become masked by the divisions between groups and the ability of the national system to divide and rule, which dates from the days of 'divide and rule' in the Empire. White supremacy still rules although perhaps less obviously than it used to—and 'white in this instance includes' some black and brown faces (see *Brown Sahibs* by Vittachi 1962). And all this while there is no movement as such within the mental health fraternity that is committed to improve the quality of services for black and other racialised groups in the UK. Race (some of us say) is 'off the agenda'.

7.4 Black People in White-Dominated Systems

The involvement of BME people in condoning institutional racism (often not realising the fact) should not surprise anyone who has read *Black Skin, White Masks* (Fanon 1952) or Fanon's later paper 'Racism and Culture', included in the book *Toward the African Revolution* (Fanon 1967). The position of black psychiatrists and psychologists, like that of many other black professionals, is complicated by the fact that they work in a system that is institutionally racist. Quite apart from problems in getting jobs in the system (see earlier 'Employment in the mental health system') progress up the career ladder often depends on the degree to which individuals follow the pathway set out by the system, which involves (for example) having one's name on one or more published papers and showing that one has undertaken research.

A noticeable change that occurred in the 1990s in the UK was the increase in numbers of people from BME communities achieving positions of some influence as professionals in the field of psychiatry—and less so in clinical psychology. In my experience, most black psychiatrists in the UK appreciate that racism is a fact in mental health services and are aware of the racist injustices faced by black users of those services, but many do not see any way in which their speaking out can alleviate this state of affairs. It is often the case that black and Asian professionals discuss these issues amongst themselves but are reluctant to do so when their white colleagues are present. Others have so internalised the culture of psychiatry—including its racist aspects—that they fail to see any problem.

Clearly, having black and Asian people in higher echelons of the professions or in positions of some prominence is very important for various reasons, not least because they may well be in a good position to understand and even counteract the effects of racism in service provision. However, this very fact sometimes hinders progress: too often, token appointments of BME people are made with the expectation that they do not rock the boat by pointing to racist practice. Also, there could be a slowing down of change and in redressing racial inequalities in mental health services by virtue of the fact that the system has black people on its staff. The very presence of BME professionals on the staff of a mental health system is frequently used as a reason for claiming that the system is not racist. When a black nurse was pursuing a grievance procedure on the basis of racism in a hospital I was working at in the early 1990s, the hospital authorities claimed that they (as an institution) could not be racist because they had a black consultant on its staff. I was told that when in the 1990s, black consultants were appointed to the prestigious Institute of Psychiatry (IOP) in London, one of these appointees was told by a senior member of staff not to follow the example set by 'Dr. Fernando'!

The story (above) resonates with another I heard recently. The National Association for Mental Health has been known as MIND for many years; and I was quite active in supporting Mind during the 1980s and 1990s, serving for some time on its Council of Management. Judi Clements, the Director/Chief Executive Officer of Mind for many years in the 1990s, told me that in the late 1990s, the then President of the Royal College of Psychiatrists (RCP) had invited her to lunch to warn her about me. He had told her that I was a very dangerous person that MIND needed to be careful of. Judi knew my work and knew me as a person, so this warning was not taken seriously by her, except to make her wonder what the RCP was so worried about. This was at the time when a Labour government had been elected and the new junior minister was a black man who had (before his election and in speeches soon after) been critical of racism in the mental health services. So, maybe the RCP was afraid at the time of a confrontation with a government minister over race matters. Unfortunately the hope that this minister would actually keep his promises to rid the psychiatric system of racism was ill founded (see references to some of these events under 'Institutional racism in the Department of Health (DOH)', earlier in this chapter).

Many BME psychiatrists and psychologists who appreciate the extent to which the clinical services are institutionally racist opt to desist from trying to change the racist institutions they work in; but some make a stand even at the risk of their own careers. However, the difficulties of being involved in this way are not something to be taken lightly. Explorations of my own experience in this field while a member of the Mental Health Act Commission is briefly described in Chap. 6 under 'Illustrations of institutional racism' and also discussed in a journal paper 'Black people in white institutions' (Fernando 1996). The conclusions of that paper state that:

> black people who wish to fight racism must become involved in British institutions, pushing their way forward if necessary, but once they are involved, they must (in order of priority): (a) be constantly vigilant and not mistake words for actions; (b) push the frontiers of anti-racism as far as possible, making alliances with anyone who wishes to co-operate; and (c) be prepared to confront racism thoughtfully and realistically. (1996, p. 152)

7.5 How Whiteness Operates

The idea of 'whiteness', more than 'blackness' (often contrasted with whiteness but used in a derogatory way until the Black Power movement in the USA turned it around to some extent), is pervasive throughout the world, indicating what is good, pure or desirable, and determining much of the power exerted by racism in many societies and in many fields of thought and experience (see 'Power of racism' in Chap. 3). But whiteness must be seen against the background of the development of the concept of 'race' itself (and the racism that followed) in association with the notion of white supremacy—the background being that of race-slavery in America and the colonisation of many parts of the world by white Europeans (see 'Exploration, colonialism, race-slavery' in Chap. 2) Then came racialisation; people who were not racialised were seen as the 'norm', against which these 'others'—the racial 'Other'—were perceived. But whiteness, like blackness, is not something that is once and for all—'it has always been a process of becoming' (Gabriel 1998, p. 49); it is still being made and remade, just as racialisation is being made and remade (see 'Racialisation' in Chap. 5).

European nations had experience of meeting and trading with people from other continents from pre-medieval times, long before the European empires. Stories were brought back to Europe about the seemingly primitive people in exotic countries, many drawing clear distinctions between people from Europe and these 'Other' people and their cultures: *they* were different to *us*. 'The pre-occupation with heathenism, promiscuity and cultural difference became the focus of intense curiosity and fantasy on the part of English [and other European] traders, travellers and writers who simultaneously re-affirmed a white English [or French, Dutch and so on] ethnocentrism inscribed with its baggage of aesthetic and religious values' (Gabriel 1998, p. 43). The elaboration of American notions of whiteness over the years has incorporated not just fantasies prevalent in Europe but the highly charged and often brutal master–slave (or master–serf) relationships between people occupying the same geographical space and interacting at various levels as individual human beings, forming relationships and so on—in a mixture of subjective and objective forms of association. So the whiteness that developed in the USA (and most of the rest of North America) was somewhat different to that which we see in most of Europe, including the UK.

7.6 Privilege and Power

The 'underlying core dimension of whiteness [in both the USA and Europe] is its capacity to secure white privilege' (Bhattacharyya et al. 2002, p. 26)—and privilege leads to power. This means unearned advantages over others in a variety of situations—from being served in a shop or restaurant to not being pulled over by police while driving. The advantages appear to be non-racial and so are assumed to be deserved—and are seen by their recipients as perfectly natural; all just by virtue of being—or appearing to be—white.

From the beginning, white privilege has promoted assumptions of, and been the basis for, white supremacy in action—politically, economically and socially. And privilege leads to power (white power) exercised silently without the person concerned having to seek it or even be aware of it (although it may be evident to others). Such power may enable a (white) person to have their views (however ridiculous they may be) taken seriously, gain political advantages and so on; and this could lead eventually to overt racism in political movements based on white power. For example, the Nazis Party (in 1930s Germany) had anti-Semitism and other forms of racism at its core. The interplay between the power wielded in clinical systems of psychology and psychiatry and (the power of) racism results in psychiatric/psychological power being felt by racialised groups as race power—in other words, the use of psychiatric power coalesces with racism (see also 'Discrimination, diagnosis and power' in Chap. 5). But power is not just something that can be recognised in actions; sometimes its effects can only be identified in terms of subjectivity. Here is an example of how it can affect people's feelings: when I was in psychiatric practice, a sentiment I heard from some of the black patients who were sectioned (compulsorily detained) and given forcible medication, was that 'it's like slavery', meaning that the psychiatry they experienced *felt* like the experience of slavery (or what they envisaged it to be).

In the clinical field, power determines and perpetuates racism through taking on the racist stereotypes available in Western culture. Take for example the production of knowledge in the course of reporting research findings by publishing papers in reputable journals. The peer-review process often controls the publication of ideas which are out of line with mainstream perspectives (Howitt 1991) and this goes for editing too. In discussing publication of cross-cultural research, Howitt and Owusu-Bempah (1994) write: 'It is not only the collection of data on black people which must be authenticated by white people, but also the language in which black people's reality or experiences are expressed' (p. 136). I can attest to academic papers being rejected on the grounds of language being 'polemic' or 'political' when the papers have been merely critical of racist practices or refer to racism in the history of psychiatry. In this regard, it is well to note that even when the editors and peer reviewers are not obviously 'white', the white norms tend to apply and their actions are nearly always in keeping with what is expected. Something that is important for the themes being pursued in this book is that the concept of whiteness (and its associated privileges and power) has/have extended to apply not just to individuals but to fields of activity and/or study, such as education or knowledge generation (whiteness of education or knowledge) or specific academic or practical disciplines. Thus one could refer to the whiteness of the system of knowledge in psychology or psychiatry. Challenging system of white knowledge is to challenge institutional racism in the 'psy' disciplines.

7.7 White Knowledge

Many universities in the UK today advertise themselves to prospective students as 'global universities', implying that the courses they provide are drawn from sources of knowledge from all over the world and are suitable for students from all over the world. Yet this is far from the case. Many university courses and training schemes in the social sciences, history, philosophy and human sciences, including those in the 'psy' disciplines, overlook the limitations of the knowledge sources they draw from. In 2015, students at University College London (UCL) raised the question of the relevance of their curricula, which were derived largely from white knowledge, by making a film, *Why Is My Curriculum White* (UCL 2015). This led to a campaign among students in several British universities demanding changes to curricula (El Magd 2016). Introducing the theme of white knowledge in '8 Reasons the Curriculum is White', the Collective, UCL (of students) (2015) state:

> Whiteness is powerful because it's unmarked and normalised. In the same way people racialised as white are considered the default human being, to which people associated with all other racialised categories are compared, much of what we consider to be the production of knowledge is, in fact, the reproduction of an ideology: whiteness.

White knowledge is by definition *biased* knowledge, often imbued with racist ideologies and assumptions, and even more importantly, the marginalisation, or even total exclusion, of knowledge derived from non-European (non-white) sources. Whiteness of knowledge evokes cultural imperialism derived from the days of Empire; and its perpetuation by academia and professionals in powerful positions is a form of institutional racism. And this is actively perpetuated by forms of peer-reviewing of papers submitted to journals or activities in support of corporations (see above under 'Privilege and power'). In a wider context, a circle of dominance is created—white knowledge interacting with capitalist forces, often neocolonialist in nature (see Fernando 2014). One way (indirect though it may be) of counteracting racism is to counteract white power, for example by ensuring the *universality* of knowledge sources and promoting the knowledge of how forces of white privilege, power and racism have influenced the production of knowledge (see Willinsky 1998).

Knowledge about the histories of professional practice and the theories on which they are based play an important part in education (in universities and schools), and as the source for training in mental health for the clinical disciplines of psychology and psychiatry. When knowledge is predominantly white, racism gets embedded in what is taught in terms of theories about the 'mind' and human interactions, ideas of illness and health, and so on; and in the curricula that trainees in the clinical disciplines are exposed to. Thus in the case of psychology and psychiatry (and other disciplines informed by these 'psy' disciplines), trained professionals are handicapped when it comes to working in ethnically and racially diverse societies, such as those in many parts of Europe and North America. In an age when racism is increasingly difficult to counteract because of the subtleties in its implementation and the resistance of white establishments against considering changing their traditional ways of working, the way forward in eradicating racism may lie in focusing on associated forces, namely those allied to whiteness. Also the very process of training is adversely affected. Most groups of trainees in the West are mixed in that they include people from various cultural and racial backgrounds. If the knowledge used during the training sessions implies (or even sometimes explicitly states) that systems created outside 'the West'—by 'Other' cultures—are 'primitive' or in some way below par, not worth considering, trainees who are seen as coming from these other cultures would be identified as actually *being* the 'Other', and naturally seen as primitive.

Some years ago I was talking to a trainee psychologist working as a CBT therapist under the supervision of a senior clinical psychologist. This young woman, a British-born Muslim person whose parents had migrated to the UK from Bangladesh, was employed to work with Muslim women allegedly suffering from depression. We talked about her work. She told me that most of the women she was asked to see attributed the feelings of depression and anxiety they experienced to the 'jinn' (a mythical supernatural influence, usually interpreted metaphorically). Her supervisor (a well-known senior clinical psychologist) had told her that such superstition represented faulty thinking and that she (the CBT therapist) should try to enable the Muslim clients to correct their thinking patterns. The trainee CBT therapist told me that she (the trainee) too believed that many of the untoward feelings were due to the jinn, and so wondered whether she herself was unhinged in some way and could not become a proper therapist.

References

Anwar, M., & Ali, A. (1987). *Overseas doctors: Experience and expectations. A research study*. London: Commission for Racial Equality.

Berman, G., & Dar, A. (2013). *Prison population statistics*. London: House of Commons. Retrieved March 10, 2015 from http://www.parliament.uk/business/publications/research/briefing-papers/SN04334/prison-population-statistics.

Bhattacharyya, G., Gabriel, J., & Small, S. (2002). *Race and power: Global racism in the twenty-first century*. London: Routledge.

DOE (Department of Education). (2012). *A profile of pupil exclusions in England* (Research Report DFE-RR190). London: DOE. Retrieved March 10, 2015 from https://www.gov.uk/government/uploads/system/uploads/attachment_data/file/183498/DFE-RR190.pdf.

EHRC (Equality and Human Rights Commission) & Runnymede Trust. (2015). *From local voices to global audience: Engaging with the international convention on the elimination of all forms of racial discrimination* (2nd ed.). Retrieved on August 19, 2016 from http://runnymedetrust.org/projects-and-publications/europe/cerd.html.

EHRC (Equality and Human Rights Commission). (2016). *Race rights in the UK*. Submission to the UN committee on the elimination of racial discrimination in advance of the public examination of ICERD. London: EHRC. Retrieved August 19, 2016 from https://www.equalityhumanrights.com/en/our-human-rights-work/monitoring-and-promoting-un-treaties/international-convention-elimination-all.

El Magd, N. A. (2016). *Why is my curriculum white? Decolonising the Academy*. (Blog in *NUS Connect*, 9 February). Retrieved on April 10, 2017 from http://www.nusconnect.org.uk/articles/why-is-my-curriculum-white-decolonising-the-academy.

Fanon, F. (1952). *Peau noire, masques blancs*. (Editions de Seuil, Paris). *Black Skin, White Masks* (C. L. Markmann, Trans.). New York: Grove Press, 1967.

Fanon, F. (1967). Racism and culture (text of Franz Fanon's speech before the first Congress of Negro Writers and Artists in Paris, September 1965, and published in the special issue of *Présence Africaine*, June–November, 1956). In F. Maspero (Ed.), *Toward the African revolution: Political essays* (H. Chevalier, Trans.), (pp. 31–44). New York: Grove Press.

Fearon, P., Kirkbride, J. B., Morgan, C., Dazzan, P., Morgan, K., Lloyd, T., et al. (2006). Incidence of schizophrenia and other psychosis in ethnic minority groups: Results from the MRC AESOP Study, *Psychological Medicine, 26*, 1541–1550.

Fernando, S. (1996). Black people working in white institutions: Lessons from personal experience. *Human Systems: The Journal of Systemic Consultation and Management, 7*(2–3), 143–154.

Fernando, S. (1998 November/December). Open letter to Frank Dobson. *Openmind, 94,* 15.

Fernando, S. (2014). *Mental health worldwide: Culture, globalization and development.* Basingstoke: Palgrave Macmillan.

Gabriel, J. (1998). *Whitewash: Racialized politics and the media.* London: Routledge.

Howitt, D. (1991). *Concerning psychology: Psychology applied to social issues.* Milton Keynes: Open University Press.

Howitt, D., & Owusu-Bempah, J. (1994). *The racism of psychology: Time for a change.* Hemel Hempstead: Harvester-Wheatsheaf.

Inyama, C. (2009). Race relations, mental health and human rights—The legal framework. In S. Fernando & F. Keating (Eds.), *Mental health in a multi-ethnic society* (pp. 27–41). London: Routledge.

Jolliff, T. (2016). The NHS is moving backwards on race equality. *The Guardian,* 31 May. Retrieved on August 22, 2016 from https://www.theguardian.com/healthcare-network/2016/may/31/nhs-moving-backwards-race-equality.

Kline, R. (2014). *The 'Snowy White Peaks' of the NHS: A Survey of Discrimination in Governance and Leadership and the Potential Impact on Patient Care in London and England.* (Report held at Middlesex University Research Repository). Retrieved March 26, 2017 from http://eprints.mdx.ac.uk/13201/.

NIMHE (National Institute for Mental Health in England). (2003). *Inside outside: Improving mental health services for black and minority ethnic communities in England.* London: Department of Health. Retrieved October 10, 2016 from http://webarchive.nationalarchives.gov.uk/+/www.dh.gov.uk/en/Publicationsandstatistics/Publications/PublicationsPolicyAndGuidance/DH_4084558.

Race Relations Act. (1976). London: Her Majesty's Stationery Office.

Sashidharan, S. P. (2006). Personal communication to author.

Stop Watch. (2017). *Metropolitan police: How many searches do the police do?* Retrieved April 1, 2017 from http://www.stop-watch.org/your-area/area/metropolitan.

Taylor, R. (2014). Section 60: A most draconian stop-and-search law that plays to police prejudice, *The Guardian,* 6 February. Retrieved February 16, 2017 from https://www.theguardian.com/commentisfree/2014/feb/06/section-60-draconian-stop-and-search-police-prejudice.

The National Archives. (1994). Powers of police to stop and search—Section 60. *Criminal Justice and Public Order Act 1994.* (The official home of UK legislation). Retrieved February 9, 2017 from http://www.legislation.gov.uk/ukpga/1994/33/section/60.

TSO (The Stationery Office). (2010). *Equality Act (2010)*. Norwich: TSO. Retrieved November 20, 2016 from http://www.legislation.gov.uk/ukpga/2010/15/contents.

UCL (University College London). (2015). Why is my curriculum white? YouTube. Retrieved September 28, 2016 from https://www.youtube.com/watch?v=Dscx4h2l-Pk.

UN (United Nations). (1965). *United nations international convention on the elimination of all forms of racial discrimination (CERD)*. Geneva Office of the High Commissioner, United Nations Human Rights. Retrieved February 10, 2017 from http://www.ohchr.org/EN/ProfessionalInterest/Pages/CERD.aspx.

Vittachi, T. (1962). *The Brown Sahib*. London: Andre Deutsch.

Weigel, M. (2016). Political correctness: How the right invented a phantom enemy. *The Guardian*, 30 November 2016. Retrieved December 10, 2016 from https://www.theguardian.com/us-news/2016/nov/30/political-correctness-how-the-right-invented-phantom-enemy-donald-trump.

Willinsky, J. (1998). *Learning to divide the world: Education at empire's end*. Minneapolis: University of Minnesota Press.

CHAPTER 8

Racism Post-9/11

In September 2001, militants claiming to be inspired by a politicised notion of jihad launched an attack on the Twin Towers in New York on 11 September ('9/11'). I referred in Chap. 4 to the end of WWII as a cultural-political watershed in Europe and North America, when hopes were raised of the West turning its back on racism and supporting human rights. However, 9/11 (2001) was another turning point—a turning back as far as the struggle against racism is concerned—leading to an uncertain future not only for multiculturalism in Western societies but for the safety of its racialised groups, often more like diasporas rather than minority ethnic groups (see 'Diasporic identities, nationalisms and multiculturalism' later in this chapter). Yet the election in 2008 of Barack Obama as president of the most powerful country in the world, the place where European racism itself was born in the age of race-slavery (see 'Exploration, colonialism, race-slavery' in Chap. 2), seemed to signal the hope of progress towards a racism-free world—a hope that was dashed by the advent of Trump in 2016, which is considered in Chap. 9 (Racism Post-Trump Post-Brexit). Pankaj Mishra (2017) sees the present period in Europe and North America as the 'age of anger'(book title), with 'more West-versus-the Rest thinking since 9/11' (p. 17). The next few paragraphs focus on the UK to explore these issues, after a paragraph considering ways in which racialised groups themselves have changed recently.

© The Author(s) 2017
S. Fernando, *Institutional Racism in Psychiatry and Clinical Psychology*, Contemporary Black History,
DOI 10.1007/978-3-319-62728-1_8

8.1 Diasporic Identities, Nationalisms and Multiculturalism

In the UK, relations between black people and white people (to use the terms I got used to in the 1960s) have undergone marked changes; 'race' itself is no longer black and white (see 'Racialisation' in Chap. 5) but yet, as the French epigram attributed to French writer Jean-Baptise Karr (1866) goes, 'plus ça change, plus c'est la même chose'—translated as 'the more things change, the more they stay the same.' Today (mid-2017) in Britain, the racial 'Other' (in the eyes of white supremacy thinking) refer to themselves as 'minority ethnic' or more often as part of a BME people, South-Asian, African-Caribbean, Chinese, African, Turkish and others who carry a 'foreign' heritage—they are Britain's racialised groups. Predominantly, most BME people see themselves as British—never 'English', although in Scotland and Wales they seem comfortable in being Scottish or Welsh. On official forms and in day-to-day conversations, most BME (racialised) people settle for hyphenated 'ethnic' designations— African-Caribbean, British-Asian and so on. But get a bit deeper into identity and many think in global terms, although much of this is more 'imaginary' than 'real' (see Anderson 1991 for discussion of imagined communities), something that is made possible through the ease of communication (for example) by internet, mobile phone and Skype. One of the consequences of globalised thinking and ease of communication is that ethnic minorities in some Western countries have become 'diasporas' with multiple identities and allegiances, a diversity of world views and expectations, and a variety of lifestyles and behaviours (Bauman 2011). This is particularly true for people living in the major cities of the UK, especially London.

There is considerable confusion today in Britain about what constitutes ethnicity, culture, 'race' and even 'nationality', resulting sometimes in controversial discussions about 'identity' and 'belonging'—issues around 'us and them' and about what human rights mean in practical terms (see, for example discussions in Hall and du Guy 1996). Most societies that we define as 'multicultural' are in today's reality composed of people with a diversity of backgrounds in terms of heritage, religious persuasion, parental identities, world views, and so on; and it is difficult to subsume any one group as differing from another in terms of a generalised, specific 'culture'. In reality, there are few clear boundaries between cultural groups and an increasing *hybridity* (mixed nature) of a culture of *individuals* (see Bhabha 1994; Pieterse 2007, 2009). So many of the groups we refer to loosely as 'cultural' are, on the whole, *racialised* groups, especially if their physical appearance or skin colour is not typical of the 'white English' (see discussion of the 'muddle between race and culture' in Fernando 2010, pp. 12–13). What has happened in both the UK and the USA is that disjunctions—even enmities—have arisen between different racialised groups and sometimes within them; for example, even Obama was sometimes said to be 'not black enough' to be called 'black' (see Toure 2011).

Although we still refer to multicultural societies, the concept of 'culture' as an overriding, overarching quality adequate for the task of defining differences between racial or ethnic groups of people, and in particular defining differences between individuals, has lost its power and now seems mostly an irrelevance. In other words, 'culture' (of an individual or group of individuals) is today seen as 'something that cannot be clearly defined … something living, dynamic and changing—a flexible system of values and worldviews that people live by, and through which they define and negotiate their lives' (Fernando 2010, p. 10). Yet 'race' continues, although now seen more in a social sense rather than a biological one, taking on board historical and political viewpoints in a context of power relationships (see discussion in Fernando 2010, pp. 7–9 and 'Definitions of racism and race' in Chap. 4). All this impinges on how mental health and ill-health are conceptualised, on what constitutes normality and madness—especially madness requiring compulsory 'treatment'—and more generally on how, and to what extent, mental health and social care services should—or could— promote the health and well-being of people. And, this state of affairs is aggravated by the legacy of cultural arrogance and institutionalised racism derived from an era when Britain, itself then largely uni-cultural, dominated a diverse number of racial/ethnic/cultural/national groups of people (see section 'Exploration, colonialism, race-slavery' in Chap. 2).

The nature of what in the UK is called 'multiculturalism' and the way racism plays out are changing—for example with tensions between Asian and Chinese groups on the one hand, and black groups on the other, complicated by Islamophobia; and anti-Semitism becoming muddled with opposition to the policies of right-wing Israeli governments (Ringler 2016). These complex interactions and intersections augur a hard time ahead for racialised groups in the West and an even harder time for anti-racism— meaning effective action against racism. And, not surprisingly, unlike in the 1970s and 1980s, there seems to be little unity in the decade that began with 2010 between ethnic/diasporic groups in fighting racism. This is particularly evident in the UK because in the USA, black African Americans still seem able to come together whenever there is a particular event or series of events—as they do in the Black Lives Matter (BLM) movement (Sidner and Simon 2015; Dabashi 2016).

8.2 Obama Years

An event that seemed at the time to represent a major shift in the race scene in the USA was the election (in November 2008) of Barack Obama, the very first black President of the USA. There were great expectations and there was talk of the USA becoming 'post racial' or 'post-race'. Obama himself campaigned on a theme of *hope*—of change; and many black African Americans cautiously hoped that at least the overt racism of police oppression might diminish with a black man occupying the White House. A narrative grew up during the Obama years—one that seemed to cross the Atlantic to be applied to the UK, where black and brown faces began to appear in positions of importance (for example in the House of Lords)—that the West was 'post-race', at least past its worst time as far as racism is concerned. Hope was high that Obama would devote his second term as president to tackling race issues; but his apparent powerlessness in the face of the rise of overt racism towards the end of his presidency, as evidenced by the rise in killings of black people by (predominantly white) police officers during 2015 and 2016 (Mapping Police Violence 2016; The Counted 2016; Lowery 2017) put paid to that hope. The notion of a post-race USA was seriously contested (for example, by Bonilla-Silva 2014); and Hero et al. (2013) suggested that 'even if one seeks to give the postracial interpretation its due—which we interpret to be the relatively narrow claim that the outcome of the 2008 presidential election demonstrates a trend towards a postracial reality—evidence that contradicts it is as substantial as that which supports it' (2013, p. 44). Today, the Obama years are increasingly seen as having had no real effect on levels of racism in the USA or anywhere else. And increasing numbers of writers argue that the experience of black people contradicts the post-race narrative.

In a *New York Times* bestseller, *The New Jim Crow: Mass Incarceration in the Age of Colorblindness,* Michelle Alexander (2014), civil rights lawyer and legal scholar, outlines a few facts that run counter to the 'post-race' narrative. 'More African American adults are under correctional control today—in prison or jail, on probation or parole—than were enslaved in

1850, a decade before the Civil War began. ... Thousands of black men have disappeared into jails, locked away for drug crimes that are largely ignored when committed by whites. ... More [black men] are disenfranchised [due to felon disenfranchisement laws] today than in 1870, the year the Fifteenth Amendment was ratified prohibiting laws that explicitly deny the right to vote on the basis of race. ... Young black men today may be just as likely to suffer discrimination in employment, housing, public benefits, and jury service as a black man in the Jim Crow era—discrimination that is perfectly legal, because it is based on one's criminal record' (pp. 180–181). Alexander makes the point that '[p]eople who have been convicted of felonies almost never truly re-enter the society they inhabited prior to their conviction. Instead, they enter a separate society, a world hidden from public view, governed by a set of oppressive and discriminatory ruled and laws that do not apply to everyone else. They become members of an undercaste—an enormous population of predominantly black and brown people who, because of the drug war, are denied basic rights and privileges of American citizenship and are permanently relegated to an inferior status' (pp. 186–187).

I first heard the phrase 'post-race' in the UK in the early 2010s. It was uttered by a representative of the Department of Health (DOH) at a meeting when the issue of overrepresentation of black people as 'schizophrenic' was being discussed (see 'Explanations for "schizophrenia" in black people' and 'Racialisation of the schizophrenia diagnosis' in Chap. 5). The message was that the problem was for black people to resolve—it was to do with their culture. In the mental health field, the talk among some of the old hands in anti-racism work is to question why 'race' is off the agenda. The number of projects in the voluntary sector (sometimes called the 'third sector'), which were important in the 1980s and 1990s in supporting black people with social and psychological problems (see 'The black voluntary sector' in Chap. 6 and Fernando 2003) has declined as a result of loss of support from funding agencies, most of which, in the UK, are state-controlled. Inequalities that represent racism in the mental health services (see 'Ethnic issues in mental health services' in Chap. 5) are worse than ever. And on top of all this, Islamophobia is opening up in the UK, exemplified by the way legislation supposed to prevent the radicalisation of young Muslim people is being implemented (see below).

Obama's election to the presidency in the USA was celebrated in the UK, not least by black and minority ethnic (BME) people involved in providing services in mental health. As in 2003 when the report *Inside Outside*

(NIMHE 2003) was published (see 'Government action' in Chap. 6), there was a feeling of hope for change but this did not last long. Instead of hope being fulfilled by actions, what happened was that society at large seemed to assume that change had already occurred—as perhaps the Nobel Prize committee thought when they awarded Obama the Peace Prize before he had had time to achieve anything significant. In both the USA and UK, the impression being fostered that societal racism declined during the latter part of the twentieth and early years of the twenty-first century is unsustainable; and it was completely exploded with the arrival of the successor to Obama and the events surrounding the decision taken by the UK to break off from the European Union (see 'Rise of the political right', 'Civil unrest' and 'Islamophobia' later this chapter; and Chap. 9).

8.3 Rise of the Political Right

The rise of the American right since the mid-1970s (Hutton 2001) exploded in 2016 with the election of Donald Trump as President of the USA in November 2016. Meanwhile in the UK too, the political right had been flexing its muscle since 9/11, resulting in successive governments appeasing right-wing parties—for example by restrictions on immigration and the right of East European nationals to obtain social benefits in the UK (Wintour 2004). In this context, anti-terrorist legislation limiting human rights rushed through in the UK as events in the aftermath of 9/11 took a distinctly racist turn, which meant that brown-skinned and black-skinned people were targeted when they 'looked like' Muslims. It is in this context that overt hostility against Muslims, which had been evident throughout the late 1990s (and had been termed 'Islamophobia' by the Runnymede Trust 1997) worsened. Muslims became racialised (see 'Racialisation' in Chap. 5). In the next few years, 'hate crime'—crimes against the person motivated by hatred against a group, or supposed group, that is racialised—showed a marked increase in both the UK and the USA, seemingly released by the notion that there was a state of war in the West, popularly seen as being against anyone perceived as 'foreign'. Statistics quoted by Seymour (2010) show that the number of racist incidents in England and Wales rose by 12% in 2003–2005 and, in 2005–2007, the number rose again by 28%, and that in Scotland, the number of racial incidents recorded per year rose from 4519 in 2004–2005 to 5243 in 2007–2008; and the UK's Crown Prosecution Service reports that the number of defendants received for racist incidents in England and Wales has risen year on year since 1999–2000—the number in

2006–2007 was almost four times the number in 1999–2000. A similar rise in overt racism has been evident in the USA especially in the lead-up to and after the election of Donald Trump as president in November 2016 (Lichtblau 2015; Metta 2015, 2016; Moore 2016).

About five months before 9/11, violence broke out in three northern cities in England, Bradford, Burnley and Oldham, in which both South Asian youth and white people were separately involved in skirmishes with police. Although dubbed 'race' riots', the well-respected *Observer* newspaper concluded in an editorial (2001) that the disturbances were a result of deliberate activities of the racist and fascist British National Party. But it was more than just that. Local economies in northern cities which had received many immigrants from Asia, predominantly Muslims from Pakistan and Bangladesh, had been decimated in the 1980s and 1990s, resulting in unemployment; for years local council-housing policies had resulted in the dumping of [South-]Asians in 'sink estates' (Phillips 2001); and white parents had been withdrawing their children from schools which they (the white parents) thought had too many Asian children (Seymour 2010); so that many schools (in the areas that the riots affected) were almost exclusively either white or South Asian. The ethnic segregation that had occurred as a result of state-sponsored policies was blamed by right-wing political movements on immigration into the region by Asians, predominantly Muslims. Unfortunately, the reports on the official inquiries into the disturbances in Bradford, Burnley and Oldham did not appear for nearly six months—by which time the 'race' context had changed dramatically as a result of 9/11—see below.

Although the main report on the 2001 disturbances, the Cantle report (Independent Review Team 2001,) recommended strategies to promote community cohesion between all communities in the areas affected, its message was drowned in the media by some extraordinary statements made by the Secretary of State for Home Affairs, David Blunkett, in which (as a renowned journalist put it): 'Going straight for the extreme, he listed all the worst aspects of religions and customs [of ethnic minorities] as if these things were everyday occurrences. ... And nothing about the white community and its obligation to reconsider the changed nature of Britishness' (Toynbee 2001, p. 16). Two other reports on the 2001 riots (Oldham Independent Panel 2001 and Burnley Task Force 2001) took a stance of highlighting: the employment discrimination against British Asians; the need to ensure racially mixed housing schemes; the fact that racial tension is exploited by organised racists; poverty; and entrenched racist views in the society. However, what was taken up by the media and government ministers was Blunkett's approach (of blaming Asians), re-enforced soon afterwards by an announcement that the government was considering an oath of allegiance for immigrants (Home Office 2002) and issuing (what Blunkett called) 'advice' to Asian people not to seek marriage partners from the Asian subcontinent (Blackstock 2002).

The change in official approach towards race issues from that of David Blunkett's predecessor at the Home Office, Jack Straw (see 'The Macpherson report' in Chap. 6) could not have been greater. The Home Office (2002) report *Secure Border, Safe Haven; Integration with Diversity on Modern Britain* clearly moved government policy away from the need to tackle institutional racism (supported only three years earlier by the same Labour government) towards 'the need to foster and renew the social fabric of our communities, and rebuild a sense of common citizenship' (p. 10). While the struggle against racism was seen in 1999 in the aftermath of the Macpherson report as one for British society as a whole, less than three years later a 'them and us' approach had again been introduced, placing the onus for change (and hence the blame for social problems) on people seen as 'migrants'—the outsider. And indeed racism has got steadily worse since then. Racism expressed in cultural language graduated into racism expressed in the language of religion or creed (as well as older ones of culture, biology, appearance and so on). And sure enough, a BBC radio series on *Race, Culture and Creed* began on 26 December 2011 by discussing the proposition 'Religion not race is the major obstacle to a multicultural society' (BBC 2011)—and by religion, what was meant was 'Islam'. All this culminated in the dramatic events of 2016—the 'Leave' campaign in the UK, which blamed immigration for many of the country's social problems and sought to exit the European Union, leading to 'Brexit' (the referendum result in June, in which the 'Leave' side won) and the presidential election campaign in the USA, leading to Donald Trump's victory in November—which permitted the outpouring of sexist, racist, xenophobic and Islamophobic sentiments on both sides of the Atlantic, with the ominous threat of unashamed overt racism re-emerging. Even so-called 'political correctness' in the use of language (see under 'New racisms in the UK' in Chap. 4 for a brief reference to political correctness) was repeatedly attacked by Trump at his election meetings—to popular applause, apparently because his supporters felt that they were freed to express racist sentiments (Weigel 2016).

8.4 Civil Unrest

In 2011, Britain saw widespread riots on the streets of London, following the shooting of a black man in London by the Metropolitan Police. Unlike the case of the riots in 1981 (see section 'Black protest in the UK' in Chap. 4) when predominantly black youngsters were involved, this time the young rioters were both black and white. The government did not institute an official inquiry as it had done in the aftermath of such occurrences in the past (see earlier this chapter and Chap. 3). A few organisations did get academics and politicians together to talk about possible causes of the riots, and the BBC was one such organisation. In a discussion on a prominent BBC programme on current affairs (*Newsnight*), David Starkey, a historian used by the BBC as an expert in analysing current events in a historical context, voiced his view that the issue in the riots was 'race'—something that surprised the other members of the panel of experts because the riots included both black and white people, proportionately with the populations in the areas affected. According to Starkey, 'the whites [who rioted] have become black' ... adopting a 'destructive nihilistic gangster culture' (*BBC News Online* 2012). Although the other panellists disagreed strongly with Starkey's proposition, several commentators later supported the implication of his contention, namely that white people absorb a propensity to bad behaviour if they commune with black people. As Starkey said, quoting Enoch Powell (see 'Transformations after WWII' in Chap. 4 for reference to Powell's 'Rivers of Blood speech' in 1968): 'His prophecy was absolutely right'. This particular racist notion is the same one put forward in the 1930s by the eminent psychologist Jung, when he blamed the 'peculiarities' of white Americans on 'racial infection' from living too close to black people (Jung, 1930—see 'Mental pathology and the construction of race-linked illnesses' in Chap. 3). Starkey's contention is yet another example of the racialisation of some aspects of culture that is a part of a 'new racism based on arguments about cultural difference' described by Cohen (1992, p. 69) and referred to in 'New racisms in the UK' in Chap. 4.

8.5 Changes in the Field of Mental Health

In September 2001, soon after 9/11, I was invited to give a talk at the annual meeting of the Black Workers' Group at the Social Services Department of Hammersmith and Fulham in West London. On meeting the (then) Director of Social Services at coffee before the meeting, I introduced myself as a consultant psychiatrist and mentioned certain funding difficulties of non-governmental organisations providing mental health services for black people (mainly African-Caribbeans) living in the area. The Director's spontaneous response was that it was 'very difficult' in the light of 9/11 to provide funds for such projects. When I questioned this (the funds were for British black people, not enemies of 'the West' or even Muslims), the Director quickly moved away from me, apparently realising the implications of what he had just said. To him groups of people seen as 'alien' to British society—the 'Other', the black—in the post-9/11 context were thought of as unacceptable in some way since 'we' were being attacked by 'them'. In fact, many non-governmental organisations serving the special needs of black people in London have felt unwelcome and in some way under suspicion since September 2001.

8.6 Racist Conclusions of Psychiatric Research

A study well known in psychiatric circles in the UK is the AESOP study (Aetiology and Ethnicity in Schizophrenia and the Other Psychoses) conducted by the Institute of Psychiatry (IOP) and reported by Fearon et al. (2006). In 2009–2010, Mathew Lewin (2009) writing in *The Guardian* newspaper stated that the researchers conducting this study had found that 'the root causes' of schizophrenia in black African Caribbean people lay in 'a whole range of social factors that led to severe social isolation—people living alone, unemployment, and the vexed issue of separation from parents due to family breakdowns in the African Caribbean community amounting to a kind of "sensory deprivation"' (p. 1). Lewin reported that the Aesop study had a major influence on the DOH and that Louis Appleby, then the government's director for Mental Health (so-called 'mental health tsar'), supported Emeritus Professor at the IOP, Julian Leff (who had led the AESOP study) in describing 'an epidemic of schizophrenia' among African-Caribbean people and advocating a 'programme of social engineering, particularly to try to strengthen family structures in the African Caribbean community with a view to keeping children in stable families'. A letter to *The Guardian* signed by 22 people including prominent professors, psychiatrists and community workers pointed to the fact that the 'inherently flawed research' that had produced the Aesop study, akin to the research leading to the publication of the Moynihan report in the USA (see 'American social studies' in Chap. 4), had been used to propose a remedy that could be 'the thin edge of a socially divisive wedge' (Ferns et al. 2010): 'Figures such as those quoted have been around for over 25 years; the real issue is what causes these differential rates of diagnosis. Is it a reflection of the true incidence of "mental illness", or is it due to other factors, including medicalisation of social problems and institutional racism?' (p. 1). Unfortunately neither Professor Leff nor Professor Appleby responded to the articles in *The Guardian*, but I reckon the points made about racism were taken by people in authority because the plan for the social engineering of black families—a racist approach if ever there was one—was not heard of again.

Some of the backstory to the issue described above indicate the problems that arise in carrying out medical research using diagnoses (such as schizophrenia) that have no scientific validity especially when they are used cross-culturally (for discussion of the diagnosis of schizophrenia in black people see 'Ethnic issues in mental health services' and 'Racialisation of the schizophrenia diagnosis' in Chap. 5)—how easily the conclusions arrived at could be affected by institutional racism. The danger is that such research may merely exacerbate racist perceptions of black people and promote the very stereotypes that result in mental health services not providing adequate and appropriate services for its black and other minority ethnic service users. In 1997–1998 I happened to learn of the research plan that eventually became the AESOP study and noted that taxpayers' money was to fund the project. I spoke with the junior minister in charge of mental health at the time who told me that he understood why the prospective research might result in conclusions that are institutionally racist.

8.7 Racism of a Psychology Report

In November of 2014, the Division of Clinical Psychology (DCP) of the British Psychological Society (BPS) issued an academic document called *Understanding Psychosis and Schizophrenia* (Cooke 2014). It was hailed—although predominantly by white people and white organisations—as a 'groundbreaking' report that constituted a 'paradigm shift' in thinking about the use of the schizophrenia category in UK. But within days of its launch, a psychiatrist, Dr. Phil Thomas, writing in a blog on *Mad in America*, a website that is popular with clinical psychologists and users and survivors of mental health services, asked why the report did not address the concerns about the diagnosis of schizophrenia that black people in the UK had been complaining about for many years (Thomas 2014).

The backstories to Phil Thomas' blog are significant in illustrating institutional racism.

Phil Thomas, Jayasree Kalathil, a survivor researcher, Jan Wallcraft, a postdoctoral researcher at the University of Wolverhampton, and I had been associated together since 2010 in launching and maintaining a website called *Inquiry into Schizophrenia Label* (ISL), calling ourselves the 'sponsors of ISL'. Soon after the blog by Thomas (on *Mad in America*), the sponsors of ISL wrote to the editor of the DCP report complaining about its failure to address issues that black people in the UK were concerned about. (Later we met with the editor and two co-authors to discuss the issues involved.) On enquiring further, we learned that all the psychologists involved in writing the report were white people—although many psychologists from BME backgrounds were members of the DCP. Also, we heard that there had been a faculty (subdivision) of the DCP called the 'Race and Culture Faculty', composed mostly of BME psychologists, for many years; that this group had been dissolved a few months before the report was published; and none of its members had been invited to participate in writing the report nor had any been invited to its launch. When we spoke with some of these BME psychologists they told us that their faculty had been dissolved by the DCP Executive Committee without any consultation with the faculty's members and that their objections were given short shrift by the Committee. We were not surprised, then, to hear from some BME clinical psychologists of discriminatory practices that were evident (to them) in appointments and interactions at a ground level and they told us of their fear of speaking out about racism within the field of clinical psychology.

When some co-authors of the report told us (the sponsors of ISL) that race and culture issues had been addressed in the report, we read it carefully and fully. It seemed that some of the content may indicate institutionally racist sentiments. The main indications of racism were evident in the following extract from a letter (not previously published) (Fernando et al. 2014) sent by the sponsors of ISL in response to a document defending the DCP report, sent by four contributors to the report.

> Surely you do not mean that supernatural, religious and spiritual explanations are only applicable to black and ethnic minority people? Such stereotyping of the 'other' is worse than misleading. And it appears to surface in other sections of the report too. For example, the impression given in paragraph 1.3 ('Our different cultures') is that people from strange cultures—the

problematic 'Other'—believe in 'demons' and patronisingly suggests that 'we' (who?) need to take into account 'their' strange beliefs! Clearly this paragraph was written for professionals. Although most psychologists who manage to make the system are white and probably trained in western scientific psychology alone, is that how it is always going to be? We have heard of instances where the IAPT services [IAPT is an acronym for a government scheme to widen the availability of psychological treatments] supervised by psychologists fail to take on board the beliefs of people seen by them as 'black and ethnic minority', seeing their ways of thinking as inferior etc. Perhaps this is not surprising when official reports are so insensitive to say the least.

There are inaccuracies in the report too. For example, there is a sentence in paragraph 6.3 when, in referring to 'migrant groups' being excessively diagnosed with schizophrenia, it states 'even though rates in their home countries are generally similar to those in the UK' (not referenced). As you may know there is a large literature on, and a long history of, attempts to explain the well-known 'over-representation' issue (of black people being excessively diagnosed with 'schizophrenia'). The over-representation is not about 'migrants' to the UK (many recent migrants are white) but of black British people—in fact the rates are higher among black British than among the black migrants of earlier years (references available). Also the unsubstantiated statement referring to 'rates in their home countries' (presumably meaning the home countries of the parents and grandparents of black people) are incorrect. We can give you references to studies that indicate that rates of 'schizophrenia' reported in [the] West Indies are lower than those reported among white people in [the] UK and considerably lower than those among black British people. Also there was a study at the Institute of Psychiatry [IOP] that throws light on how IOP psychiatrists, compared to a black Jamaican psychiatrist, made (psychiatrically speaking) diagnoses that were different in the case of black patients, over-diagnosing 'schizophrenia'. The issue of 'over-representation' is regarded by many black service users and academics who have studied the issue as one of racism, not migration or 'culture'—something that the report *Understanding Psychosis and Schizophrenia* fails to recognise. Perhaps the incorrect information about rates of diagnosed schizophrenia and racism was obtained from the Aesop study which incidentally has been the subject of much criticism for misrepresentation and some racist conclusions it draws.

(Fernando et al. 2014)

In a paper critical of the DCP report, Kalathil and Faulkner (2015) pointed to a 'conspicuous absence in the report of any meaningful engagement with the more than 60 years of scholarship about "race", ethnicity and psychiatry/psychology, especially around the theorisation and experience of schizophrenia. This lack of engagement is also evident in the absence of professionals or service users/survivors from racialized communities in the working party and in the apparent lack of attempt to gain their views in its production' (2015, p. 22). The editor of the DCP report agreed to issue an apology for what were called 'flaws' in the report and to rewrite parts of it so that it did not reflect sentiments that may be interpreted as institutionally racist, and to rectify incorrect statements on the rates of diagnosis of schizophrenia among black British people. Unfortunately, the apology turned out to be weak and, in any case, was not circulated to those who had been sent the report—in fact it had very poor circulation (Keval 2015). We learned later that the DCP continued to publicise the original (flawed) version, maintaining a stance of not seeing the flaws in the report as being very serious; and the flawed report was launched in North America and Scandinavia in spite of our appeals for launches to be delayed until the corrected report was published. It was not until June 2017 that a re-written version was finally issued. The case described in the previous paragraph has within it several aspects of whiteness—knowledge, privilege and power and also aspects of institutional racism (see 'How whiteness operates', 'Privilege and power' and 'White knowledge' in Chap. 7). The psychology report alluded to was constructed totally on the basis of white knowledge—that is, knowledge derived from Eurocentric sources alone—and reflected a white supremacy approach (that white knowledge alone is valid). There were elements of racism in institutional processes in the DCP in that if the Faculty of Race and Culture had been operational at the time the report was published, it would have had to have been passed by the faculty (a very unlikely occurrence) before being released for publication. No BME psychologist had been told of the project to write such a report and none had been invited to attend its launch and all the co-authors approached were white people. This reflects white privilege and white power in decision-making (see 'Privilege and power' in Chap. 7). A racist agenda had been pursued probably without anyone operating it being fully aware they were involved in it, and had been implemented through the power vested in the whiteness of psychologists, and in the discipline itself.

The story about the psychology report illustrates the fact that it is not a matter of subtle (institutionally transmitted) racism being the sole problem which must be tackled—the need to oppose overt racism goes without saying—but the way forward towards eradicating racism may be to counteract assumptions of white supremacy by pointing it out to the people concerned wherever white privilege is evident, challenging white power and working on the strategies of knowledge production that have resulted in white knowledge dominating the mental health field.

8.8 Islamophobia

In 2014, Baroness Warsi, a senior Tory and (then) a minister at the British Foreign Office claimed publicly that Islamophobia 'had "passed the dinner table test" and become socially acceptable in the UK' (Hasan 2014). What Warsi meant was that, whereas racism had been gradually driven 'underground' in the decades after WWII (see 'Transformations after WWII' and 'New racisms in the UK' in Chap. 4) so that it was expressed predominantly in subtle and indirect ways, racism (as Islamophobia) was now, in the second decade post-9/11, appearing above the surface again. The doors were opening to possible public acceptance of racism—reversing the trend of the previous 50 years. Warsi was implying that the UK was in danger of sliding back to a pre-war position—even that of the 1930s. And all that happened after the election campaign leading to the vote in the referendum for the UK to leave the European Union (the so-called Brexit vote), held on 24 June 2016, confirmed the worst fears of commentators such as Baroness Warsi. The rise in racist abuse on the streets and racist 'hate crime' in the UK was described in *The Guardian* newspaper of 30 June 2016 as 'a generalised kind of racism against groups perceived not to be in that narrow category of white English' (Khaleeli 2016, p. 8).

Islamophobia, or anti-Muslim hostility, like other forms of racism did not appear in Europe and Euro-America out of the blue, although its current form may represent (and been promoted by) recent events. One could argue that 'Muslim' has become a racialised identity quite recently, perhaps since the 1980s when immigrants to Europe from predominantly Muslim countries began to be identified as a separate group and not just as 'Asians' or 'Africans'; but that happened on the basis of a long-standing underlying hostility to Muslims in European culture (or 'Christendom') which had existed for almost as long as anti-Semitism. I noted earlier (the beginning of Chap. 2) that Muslims and Jews had been identified in European culture as the alien, dangerous 'Other' well before the Middle Ages; that Christian crusaders massacred both Muslims and Jews in the aftermath of capturing Jerusalem in July 1099 (Woods 2003; Phillips 2010); and that the persecution by the Christian Spanish Inquisition in the years following the defeat of the Moors at Granada in 1492 affected both Jews and Muslims (see 'Exploration, colonialism, race-slavery' in Chap. 2).

Looked at in a context of the European history of images of Muslims, it is not surprising that, when President George Bush announced his 'war on terror', he called it a 'crusade' (White House Archives 2001) just as he referred to the invasion of Iraq in 2003 by the USA and the UK as a 'crusade' (Lears 2003). One could easily conceive 'crusade' as the Christian version of the Islamic 'jihad'. The 'clash of civilisations' as the underlying cause of many recent conflicts and wars (Huntingdon 2002) may be a catchy idea but a different approach to history could well see racism (including Islamophobia and anti-Semitism) as an equally plausible one—and one that presages a remedy (namely taking effective action to minimize racism) for avoiding major conflict in the future as well as dealing with current conflicts. Kevin Schwartz (2017), a research worker at the Library of Congress, sees Islamophobia as the re-emergence of the notion of the 'conspiratorial Muslim' that has been evident in policies of various colonial powers, quoting as examples British concern over a chronic Muslim conspiracy in India in the aftermath of the 1957 rebellion (termed a mutiny by the British colonial government); and the concerns of a Dutch civil servant in Java that Javanese Muslims could threaten Dutch rule in the East Indies.

Today in Euro-America, Islamophobia affects people seen (in the West) as 'Muslims'. The identification of the 'Muslim' today in the West (for the purpose of Islamophobia) is quite complicated. The categories of 'Muslim', 'Arab' and the outdated 'Moor' have become conflated and are all seen as 'Muslim'. Beydoun (2013) describes how in the USA '[j]udges during the [racially restrictive] Naturalization Era viewed "Arab" as synonymous with "Muslim" identity'. And, since Arabs were seen as non-white, all Muslims were denied citizenship, except some Arab Christians who were able to 'invoke the fact of their Christianity to argue that they were white' (p. 29). The Muslim person subject to Islamophobia need not be a believer in Islam at all—they just have to *look* Muslim, be thought of (by others) as Muslim, be perceived as Muslim. This given identity may be because they were born in a predominantly Muslim country and/or have brown skin (or darkish) skin colour, or dress like Muslims generally do. So in the case of Palestine, both Christians and Muslims get labelled as 'Muslim' by Euro-America, by virtue of being Palestinian (Palestine is predominantly Muslim once the illegal Jewish settlers are ignored).

Stereotypes play a role in who is seen as a Muslim and who is not. The book *Orientalism* by Edward Said, Professor of Literature at Columbia

University, stands as a classic study of the images of 'the Orient' and its peoples that have been created by Europeans over centuries and that have influenced European thinking, contributing to stereotyping and racism. In discussing 'orientalism' *now* (that is, at the time of writing, in the twentieth century), Said (1978), states: 'By a concatenation of events and circumstances [since WWII], the Semitic myth [created by the West, and which combined being Oriental with being Semitic] bifurcated in the Zionist movement [that led to the political state of Israel in 1948]; one Semite [that is the Jew] went the way of Orientalism, the other, the Arab, was forced to go the way of the Oriental. ... [And] each time the concept of Arab [nowadays Muslim] national character is evoked, the myth is being employed. ... [The] powerful difference posited by the Orientalist as against the Oriental is that the former *writes about* whereas the latter *is written about*' (pp. 307–308, emphasis in the original). People referred to as Muslims today (2017) in the West are endowed, in the mind of the European, with the myths and stereotypes attributed to 'the Oriental', although they may be, for example, from China, Russia, Indonesia or Somalia and so on.

The most obvious and pervasive form of Islamophobia in the UK is that institutionalised through the legislation that has been embedded, in a convoluted fashion, in the British government's 'Prevent Strategy' (HM Government 2011)—ostensibly designed to prevent terrorism. The strategy aims to do this by identifying those people deemed vulnerable to being 'radicalised' into developing extremist ideas, in order to take action to (if possible) de-radicalise them (see Cole 2016). The operation of Prevent has been widely criticised on the grounds of its adverse effects on human rights (Singh 2016) but even more seriously for its unethical nature in drawing professional groups from the education and health sectors into the surveillance of society (Summerfield 2016), a surveillance mostly focused on Muslim people.

8.9 The 'Psy' Disciplines and Islamophobia

The 'psy' disciplines have been seriously disadvantaged by the precursors to Islamophobia—the negative attitudes towards Islamic history and culture that are clearly evident, for example, in the selectivity of historical tomes about psychiatry and psychology. Well-known books on the history of the study of madness, such as those by Scull (2015) and Porter (1987) contain very little reference to the sorts of psychiatry and psychology practised in the mental hospitals of the Arabic Empire, although these flourished for 300 years, between the tenth and thirteen centuries (see Dols 1992 and 'Limitations of knowledge' in Chap. 2); and so, current information that informs the 'psy' disciplines does not have access to the vast stores of knowledge on mental health and mental illness in Arabic writing, partly because many Arabic books were destroyed by the Spanish Inquisition after the expulsion from Spain of Jews and Muslims around the end of the fifteenth century (see 'Exploration, colonialism, race-slavery' in Chap. 2). However it should be noted that Michel Foucault, who is renowned for his critical exploration into the background to medical and psychological knowledge, although again limiting much of his work to non-Islamic European work, mentions the fact that 'the Arab world seems to have built some early hospitals specifically for the insane … [where] … the mad were treated with a certain degree of medical humanism' before the advent of the positivism of European Enlightenment thinking (Foucault 2006, p. 117).

In 2011 the UK government introduced legislation under the Channel Duty Guidance (HM Government 2015), issued under Sections 36(7) and 38(6) of the Counter-Terrorism and Security Act 2015 (TSO 2015), that affects professionals working in the field of education and health care. Through this legislation the British state has imposed a legal duty on teachers, doctors, nurses and others in education and health bodies to report to the authorities any students and patients seen to be at risk of 'extremism'. Mental health professionals, including clinical psychologists and psychiatrists as part of the workforce in the National Health Service (NHS), have a duty to refer for assessment any of their clients who may be vulnerable to radicalisation into extremism and hence (in the view of the British government) to commit acts of terrorism. Committees have been set up under police departments to examine those referred, taking whatever action may be necessary. The way the instructions are framed makes it clear that this process applies mainly, if not entirely, to Muslims; and instructors, alleged to be qualified to detect people vulnerable to radicalisation, have been training professionals, and others working in the health and education systems, in this supposed expertise of radicalisation detection.

No clear evidence has been made available (as of April 2017) on what exactly the training for detecting vulnerability to radicalisation entails, except that it appears to rely on assessment criteria known as Extremism Risk Guidance 22+ (ERG22+)—the + representing the fact that items other than the original 22 can be used in individual cases—developed by two psychologists working at the National Offenders Management Service (NOMS). The initial work carried out at NOMS is classified (which means it is a state secret) but the authors published an article in a journal in the USA (Lloyd and Dean 2015) outlining the methodology they had used for arriving at their conclusions. There is no evidence of proper scrutiny and peer review of the work, nor of its ethical clearance from the professional bodies that the authors belonged to.

An independent advocacy group, Open Society Justice Initiative, advised by academics drawn from a range of academic backgrounds, researched the impact of the work carried out under the Prevent Strategy in the health and education sectors during 2015 and 2016 (Singh 2016). Its report raised a variety of concerns about the study by the two psychologists (as published) and the way that the 22 risk factors were derived: '[I]t is unclear how these 22+ factors, apparently developed in the context of a prison population by the National Offender Management Service, with no demonstrated link with future offending, can be applied to a general population to assess vulnerability to being drawn into terrorism' (p. 39). And the report went on to state: 'Prevent operates in a climate marked by Islamophobia, in a nation that has just voted for Brexit [the referendum on 24 June 2016 produced a majority in England for leaving the European Union], and where immigration is an issue. Significantly, between July 2015 and July 2016, Islamophobic crime in Britain rose by 94%. The case studies described below raise serious questions about whether individuals targeted under Prevent would have been targeted in this manner had they not been Muslim' (p. 52).

A letter signed by more than 140 experts (mainly academics) voices concern at the lack of 'proper scrutiny or public critique' of the study on which the ERG 22 + was based (Armstrong et al. 2016); and the Royal College of Psychiatrists (RCP) has called for publication of the full ERG22 + study (RCP 2016) stating: 'Public policy cannot be based on either no evidence or a lack of transparency about evidence' (p. 7). At the time of writing, the UK government has not responded to professional calls for transparency; the psychologists involved in deriving the risk factors (on the basis of which professionals are supposed to judge vulnerability to radicalisation) have not claimed that the ERGs have any predictive validity—in fact, the psychologists Lloyd and Dean (2015) state that the ERGs 'cannot be taken as substitute for predictive validity' (p. 80). Meanwhile the government has asked professionals in education and health (including British psychologists and psychiatrists) to use these ERGs to identify people as vulnerable to radicalisation and refer them to committees set up across the country, led by police departments.

The ERGs were based on the work of clinical psychologists. Psychiatrists, too, are apparently being drawn into the radicalisation field. Significantly, three papers have appeared to date in the psychiatric literature. The data analysed in each paper is exactly the same, and was obtained from interviewing 608 Muslims. The first paper (published in March 2014 —Bhui et al. 2014b) reported no relationship between 'depression' and 'radicalisation'; but the second (published six months later—Bhui et al. 2014a) reported a finding that 'mild depression' was associated with 'radicalisation' when the data was reallocated by a method that weighted some items of the radicalisation scale used called SVPT (sympathies for violent protest and terrorism)—which itself had not been validated—by what is called a 'cluster analysis' on a so-called 'classification likelihood method' (p. 3), although there is no attempt in the paper to justify the use of this method in this sort of study. The third paper (Bhui et al. 2016) derives complicated theories on how depressive symptoms plus either the death of a relative, another major event or the subject having signed a petition, could evoke sympathies for violent protest. Incidentally the third paper states that the first paper (Bhui et al. 2014b) 'found an association between depressive symptoms and SVPT' (Bhui et al. 2016, p. 483) although the latter states clearly that no such association was found. The three papers show a tendency towards 'data dredging/data-trawling'—the dubious approach of repeated analysing of the same data using a variety of statistical approaches and combinations of data in the hope of reaching significant findings (Young and Kerr 2011)—thus raising serious doubts about the quality of the research.

The apparent manipulation of psychiatry and psychology, in the cause of security surveillance—the securitisation of health—clearly runs the risk of the 'psy' disciplines being collusive in Islamophobic racism. History may well judge the 'psy' professions harshly for getting involved in this way, if we are mindful of how these disciplines were involved in the eugenics movement in the UK and in the justification of the enslavement and oppression of Africans during plantation slavery in the USA (see 'Origins of (Western) psychology and psychiatry' in Chap. 2 and 'Mental pathology and the construction of race-linked illnesses' in Chap. 3), as well as in oppression in the (former) Soviet Union in the 1960s and apartheid South Africa (see 'Discrimination, diagnosis and power' in Chap. 5).

8.10 Conclusions

In the past 50 years, we have come some way in minimising racism and counteracting its effects in both the USA and the UK—with positive progress during the 1980s and 1990s and a slipping back since 2001. But the progress achieved is now threatened by the uncertainties of the present (mid-2017) situation, with a post-Brexit UK and with the advent of Trump in the USA—a situation that is discussed in Chap. 9. In the case of the 'psy' disciplines the outcome of efforts to address racism since the mid-1980s has been disappointing. Even today, institutional racism sits strong in the 'psy' disciplines. The universalist psychiatric/psychological doctrine that Western concepts of the mind, of illness models and of systems of treatment have a global relevance subsumes within it a distinct racist judgement of cultures and peoples. The thinking that informed race psychology within clinical psychology remains strong—only partially concealed by political correctness of language. Psychiatric diagnoses continue to carry racist undertones, echoing the diagnoses given to black people who were enslaved way back in the days of the Atlantic slave trade. Current practitioners tend to ignore the racist dimension of their disciplines and therefore little, if any, action is usually taken to counteract the effects of institutional racism in practice—and instead of systemic change white supremacist thinking allows little more than tokenistic symbols or voiced good intentions. Consequently, not only are racist traditions perpetuated, but also the racism of Western culture continues to permeate the psy disciplines in clinical practice, research and theory. The future of the 'psy' disciplines and the way in which racism may play out in that future is discussed in Chap. 9.

References

Alexander, M. (2014). The new Jim crow. *The Huffington Post blog*, 25 May 2011.

Anderson, B. (1991). *Imagined communities, reflections on the origin and spread of nationalism.* London: Verso.

Armstrong, K., Sagemen, M., et al. (2016, September 29). Anti-radicalisation strategy lacks evidence base in science. Open letter from over 140 academics. *The Guardian.* Retrieved on 1 November 2016 from https://www.theguardian.com/politics/2016/sep/29/anti-radicalisation-strategy-lacks-evidence-base-in-science.

Bauman, Z. (2011). *Culture in a liquid modern world.* Malden, MA: Polity Press.

BBC. (2001, December 26). *Race, Culture and Creed.* First Part of three part series broadcast on Radio 4. London: British Broadcasting Corporation.

BBC News Online. (2011). England riots: 'The whites have become black' says David Starkey. BBC News 13 August 2011. Retrieved on 10 October 2016 from http://www.bbc.co.uk/news/uk-14513517.

Beydoun, K. A. (2013). Between Muslim and White: The legal construction of Arab Muslim American identity. *New York University Annual Survey of American Law*, 69(1), 29–76.

Bhui, K., Everitt, B., & Jones, E. (2014a). Might depression, psychosocial adversity, and limited social assets explain vulnerability to and resistance against violent radicalisation?'. *PLOS ONE*, 9(9), e105918.

Bhui, K., Warfa, N., & Jones, E. (2014b). Is violent radicalisation associated with poverty, migration, poor self-reported health and common mental disorders?' *PLOS ONE*, 9(3), e90718.

Bhui, K., Silva, M. J., Topciu, R. A., & Jones, E. (2016). Pathways to sympathies for violent protest and terrorism. *British Journal of Psychiatry, 209*(6), 483–490.

Blackstock, C. (2002, February 8). Blunkett in clash over marriages. *The Guardian*, 2002, p. 1.

Bonilla-Silva, E. (2014). *Racism without Racists. Color-blind racism ad the persistence of racial inequality in America* (4th ed.). New York: Rowman and Littlefield.

Burnley Task Force. (2001). *Burnley Independent Task Force Report (chairman: Lord Anthony Clarke)*. Burnley: Burnley Borough Council.

Bhabha, H. K. (1994). *The location of culture*. London: Routledge.

Cohen, P. (1992). It's racism what dunnit. Hidden narratives in theories of racism. In J. Donald & A. Rattansi (Eds.), *'Race', culture and difference* (pp. 62–103). London: Sage in association with Open University.

Cole, S. (2016). Hear us before your knock prevent—we're trying to save lives' *The Guardian* 31 October 2016. Available on 2 February 2017 from https://www.theguardian.com/commentisfree/2016/oct/31/prevent-save-lives-families-child-terrorism-programme.

Cooke, A. (ed.). (2014). *Understanding psychosis and schizophrenia*. Leicester: British Psychological Society. Retrieved on 10 January 2017 from http://www.bps.org.uk/networks-and-communities/member-microsite/division-clinical-psychology/understanding-psychosis-and-schizophrenia.

Dabashi, H. (2016, September 6). Black lives matter and palestine: A historic alliance. *Aljazeera*. Retrieved on 10 January 2017 from http://www.aljazeera.com/indepth/opinion/2016/09/black-lives-matter-palestine-historic-alliance-160906074912307.html.

Dols, M. W. (1992). *Majnūn: The Madman in Medieval Islamic Society* D. E. Immisch (Ed.). Oxford: Clarendon Press.

Editorial. (2001, November 25). Inside our changing land. In *Race in Britain* (special ed.). *The Observer*, p. 1.

Fearon, P., Kirkbride, J. B., Morgan, C., Dazzan, P., Morgan, K., Lloyd, T., et al. (2006). Incidence of schizophrenia and other psychosis in ethnic minority groups: Results from the MRC AESOP Study. *Psychological Medicine, 26,* 1541–1550.
Fernando, S. (2003). *Cultural diversity, mental health and psychiatry. The struggle against racism.* Hove: Brunner-Routledge.
Fernando, S. (2010). *Mental health, race and culture* (3rd ed.). Basingstoke: Palgrave.
Fernando, S., Thomas, P., Kalathil, J., & Wallcraft, J. (2014). Response to statement: Initial response to concerns regarding "Understanding Psychosis"'. Retrieved on 27 April 2017 from http://www.sumanfernando.com/news.html.
Ferns, P., Barker, P., Beresford, P., Blackman, P., Bracken, P., Christie, Y., et al. (2010). Poor research or the attack on black people? Letter from mental health campaigners on the alleged 'epidemic' of schizophrenia among British African Caribbean groups. *The Guardian* 3 February 2010. Retrieved on 30 July 2016 from https://www.theguardian.com/society/2010/feb/03/mental-health-bme-schizophrenia-letter.
Foucault, M. (2006). *History of Madness* (Jean Khalfa ed., J. Murphy & J. Khalfa trans.). London: Routledge.
Hall, S., & du Guy, P. (Eds.). (1996). *Questions of cultural identity.* London: Sage.
Hasan, M. (2014, March 29). Baroness Warsi speaks out on Islamophobia, Richard Dawkins, Bingo Posers and 'racist' Ukip. *Huffpost Politics United Kingdom.* Retrieved on 6 August from http://www.huffingtonpost.co.uk/2014/03/26/baroness-warsi_n_5036065.htmlAccessed.
Hero, R. E., Levy, M. E., & Radcliff, B. (2013). The end of "race" as we know it? Assessing the "postracial America". Thesis in the Obama Era'. In F. C. Harris & R. C. Lieberman (Eds.), *Beyond discrimination: Inequality in a postracist era,* pp. 39–72. New York: Russell Sage Foundation.
HM Government. (2011). *Prevent Strategy* Cm 8092. London: The Stationery Office. Retrieved on 11 October 2016 from https://www.gov.uk/government/uploads/system/uploads/attachment_data/file/97976/prevent- strategy-review.pdf.
HM Government. (2015). *Channel Duty Guidance. Protecting vulnerable people from being drawn into terrorism.* Retrieved on 10 January 2017 from https://www.gov.uk/government/publications/channel-guidance.
Home Office. (2002). *Secure borders, safe haven. Integration with diversity in modern Britain.* CM 5387. London: The Stationery Office. Retrieved on 10 October 2014 from https://www.gov.uk/government/uploads/system/uploads/attachment_data/file/250926/cm5387.pdf.
Huntingdon, S. P. (2002). *The clash of civilizations: And the remaking of world order.* New York: Simon and Schuster.

Hutton, W. (2001, December 16). Words really are important. My Blunkett. *The Guardian*. Retrieved on 17 November 2016 from https://www.theguardian.com/politics/2001/dec/16/race.world.

Independent Review Team. (2001). *The cantle report: Community Cohesion: A report of the independent review team*. (Chaired by Ted Cantle). London: Home Office. Retrieved on 17 November 2016 from http://www.ttrb3.org.uk/community-cohesion-a-report-of-the-independent-review-team/.

Jung, C. G. (1930). Your Negroid and Indian behaviour. *Forum, 83*(4), 193–199.

Kalathil, J., & Faulkner, A. (2015 Jan-Feb). Racialisation and knowledge production: A critique of the report understanding psychosis and schizophrenia. *Mental Health Today* pp. 22–23. Retrieved on 18 November from https://www.academia.edu/11196191/Racialisation_and_knowledge_production_A_critique_of_the_report_Understanding_Psychosis_and_Schizophrenia.

Karr, J.-B. A. (1866). *Les Guêpes*. Paris: Michel Lévy Frères. Retrieved on 24 January 2017 from https://archive.org/details/lesgupes05karrgoog.

Keval, H. (2015). Schizophrenia and psychosis: The magical and troubling disappearance of race from the debate. *Diversity and Equality in Health and Care* 12(1), 6–8. Retrieved on 12 January 2017 from http://www.diversityhealthcare.imedpub.com.

Khaleeli, H. (2016, June 30). Campaigners and victims reporting a rise in racist abuse since the Brexit vote. Has the hatred always been there under the surface—and will this "celebratory racism" cause lasting damage? *The Guardian supplement* pp. 6–9. Retrieved on 10 April 2017 from '"A Frenzy of hatred": how to understand Brexit racism'. https://www.theguardian.com/politics/2016/jun/29/frenzy-hatred-brexit-racism-abuse-referendum-celebratory-lasting-damage.

Lears, J. (2003, March 11). How a war became a crusade. *The New York Times*. Retrieved on 17 January from http://www.nytimes.com/2003/03/11/opinion/how-a-war-became-a-crusade.html.

Lewin, M. (2009, December 9). 'Schizophrenia "epidemic" among African Caribbeans spurs prevention policy change' *Society Guardian*. Retrieved on 30 July 2016 from https://www.theguardian.com/society/2009/dec/09/african-caribbean-schizophrenia-policy.

Lichtblau, E. (2015, December 17). Crimes against Muslim Americans and Mosques Rise sharply. *New York Times on line*. Retrieved on 15 November 2016 from http://www.nytimes.com/2015/12/18/us/politics/crimes-against-muslim-americans-and-mosques-rise-sharply.html?_r=0.

Lloyd, M., & Dean, C. (2015). The development of structured guidelines for assessing risk in extremist offenders. *Journal of Threat Assessment and Management, 2*(1), 40–52.

Lowery, W. (2017, January 1). Black lives matter: Birth of a movement. *The Guardian*, 23–25. Retrieved on 17 January from https://www.

theguardian.com/us-news/2017/jan/17/black-lives-matter-birth-of-a-movement.

Mapping Police Violence. (2016). Police have killed at least 230 black people in the US in 2016. Retrieved on 11 October from http://mappingpoliceviolence.org/.

Metta, J. (2015). Racism in the US—the melting pot is boiling. *Aljazeera Opinion on line*, Retrieved on 16 November from http://www.aljazeera.com/indepth/opinion/2015/12/racism-melting-pot-boiling-151210051654391.html.

Metta, J. (2016, November 10). The foul stench of fascism in the US. *Aljazeera News on line*. Retrieved on 16 November from http://www.aljazeera.com/indepth/features/2016/11/foul-stench-fascism-161110100340474.html.

Moore, P. (2016, March). Divide on Muslim neighbourhood patrols but majority now back Muslim travel ban. *YouGov*. Retrieved on 15 November from https://today.yougov.com/news/2016/03/28/divide-muslim-neighborhood-patrols/.

Mishra, P. (2017). *Age of anger. A history of the present*. London: Allen Lane Penguin Random House.

NIMHE. (National Institute for Mental Health in England). (2003). *Inside outside. improving mental health services for black and minority ethnic communities in England*. London: Department of Health. Retrieved on 10 October 2016 from http://webarchive.nationalarchives.gov.uk/+/www.dh.gov.uk/en/Publicationsandstatistics/Publications/PublicationsPolicyAndGuidance/DH_4084558.

Oldham Independent Panel. (2001). *Oldham independent review. One Oldham One Future*. Oldham, Greater Manchester: Oldham Metropolitan Borough Council (panel chairman: D. Ritchie). Retrieved on 10 October 2016 from http://www.tedcantle.co.uk/publications/002OneOldham,OneFutureRitchie2001.pdf.

Phillips, T. (2001) White flight is enforcing segregation. The Guardian, 19 December 2001, p. 20.

Phillips, J. (2010). *Holy warriors: A modern history of the crusades*. London: Vintage.

Pieterse, J. N. (2009). *Globalization and culture global Mélange*. Lanham, MA, USA: Rowman and Littlefield.

Pieterse, J. N. (2007). *Ethnicities and global multiculture pants for an Octopus*. Lanham, MA, USA: Rowman and Littlefield.

Porter, R. (1987). *A social history of madness. Stories of the Insane*. London: Weidenfeld and Nicholson.

Ringler, S. (2016, December 12). The mistaken equivalence of anti-semitism and anti-zionism. *ReformJudiasm.org*. Available on 12 January 2017 at: http://www.reformjudaism.org/blog/2016/12/21/mistaken-equivalency-anti-semitism-and-anti-zionism.

RCP. (Royal College of Psychiatrists). (2016). *Count-terrorism and psychiatry. Position statement* PS04/16. London: RCP.

Runnymede Trust. (1997). *Islamophobia, a challenge for us all*. London: Runnymede Trust.

Schwartz, K. (2017, January 17). The Muslim as a 'Manchurian candidate. The imperial-age idea of the 'conspiratorial Muslim' has re-emerged with new force and scope. *Aljazeera News*. Retrieved on 17 January 2017 from http://www.aljazeera.com/indepth/opinion/2017/01/muslim-manchurian-candidate-170116085751234.html.

Scull, A. (2015). *Madness in civilization. A cultural history of insanity from the Bible to Freud, from the Madhouse to Modern Medicine*. London: Thames and Hudson.

Said, E. (1978). *Orientalism*. New York: Random House.

Seymour, R. (2010). The changing face of racism. *International Socialism, A quarterly review of socialist theory*, 126. Published on line. Retrieved on 10 January 2017 from http://isj.org.uk/the-changing-face-of-racism/#126seymour_6.

Sidner, S., & Simon, M. (2015, December 28). The rise of black lives matter: Trying to break the cycle of violence and silence. *CNN*. Retrieved on 10 January 2017 from http://edition.cnn.com/2015/12/28/us/black-lives-matter-evolution/.

Singh, A. (2016). *Eroding trust. The UK's Prevent Counter-Extremism Strategy in Health and Education*. New York: Open Society Foundations. Retrieved on 14 January 2017 from https://www.opensocietyfoundations.org/reports/eroding-trust-uk-s-prevent-counter-extremism-strategy-health-and-education.

Summerfield, D. (2016). Mandating doctors to attend counter-terrorism workshops is medical unethical. *British Journal of Psychiatry Bulletin*, 40(1–2), 87–88. doi:10.1192/pb.bp.115.053173.

The Counted. (2016). People killed by police in the US, recorded by *The Guardian*. Available on 10 December 2016 at https://www.theguardian.com/us-news/ng-interactive/2015/jun/01/the-counted-police-killings-us-database.

Thomas, P. (2014). *Understanding psychosis and schizophrenia? What about black people?* Retrieved 10 February 2017 from http://www.madinamerica.com/2014/12/dcpbps-report-understanding-psychosis-schizophrenia-fatally-flawed/.

TSO. (The Stationery Office). (2015). *Counter-terrorism and security Act 2015. Chapter 6*. London: TSO. Retrieved on 20 November 2016 from http://www.legislation.gov.uk/ukpga/2015/6/pdfs/ukpga_20150006_en.pdf.

Touré (2011) *Whose afraid of post-blackness now? What it means to be black*. New York: Free Press.

Toynbee, P. (2001, December 12). Religion must be removed from all functions of state. *The Guardian*. p. 18. Retrieved on 10 February 2017 from https://www.theguardian.com/society/2001/dec/12/communities.comment.

Weigel, M. (2016, November 30). Political correctness: How the right invented a phantom enemy. *The Guardian*. Retrieved on 10 December 2016 from https://www.theguardian.com/us-news/2016/nov/30/political-correctness-how-the-right-invented-phantom-enemy-donald-trump.

White House Archives. (2001, September 16). Remarks by president upon arrival. *Office of the Press Secretary*. Retrieved on 17 January 2017 from https://georgewbush-whitehouse.archives.gov/news/releases/2001/09/20010916-2.html.

Wintour, P. (2004) Blunket urged to resist migrant crackdown. The Guardian, 23 February 2004. Retrieved on 10 December 2016 from https://www.theguardian.com/politics/2004/feb/23/eu.thinktanks.

Woods, A. (2003). George W. Bush and the Crusades. In Defence of Marxism 08 May 2003. Retrieved on 17 January 2017 from https://www.marxist.com/iraq-bush-crusades080503.htm.

Young, S. S. & Karr, A. (2011). Deming, data and observational studies. *Significance*, 8(3), 116–120. Retrieved on 30 March 2017 from http://onlinelibrary.wiley.com/wol1/doi/10.1111/j.1740-9713.2011.00506.x/full.

CHAPTER 9

Racism with the Advent of Trump and After Brexit

Mid-2017 is a time of uncertainty and fear about the future—a result of the election of the USA's current president, a demagogue who campaigned on a platform of racist and xenophobic nationalism that now threatens human rights the world over (Nougayrède 2017), foreshadowed by the UK's vote (in a referendum) to break with the rest of Europe after a campaign characterised by racism and anti-immigrant sentiments (Khaleeli 2016). Further, nationalisms of a racist nature are rising in many part of the world, such as in France (see Neslen 2014) and India (see Lakshmi 2015), allied to right-wing political movements. National identities (linked to countries of origin) are evident in diasporic communities in the West (see 'Diasporic identities, nationalisms and multiculturalism' in Chap. 8). Although it is theoretically possible that these identities may give rise to ethnic nationalisms which may clash with the nationalisms allied to white supremacist thinking in the majority ethnic populations, the evidence is that major cities in Europe and the USA (where the vast majority of diasporic communities live) are a bulwark against narrow nationalisms and the rise of racism. Ishan Tharoor (2016) quotes Steve Bannon, the white nationalist ideologue who advises Trump, as grumbling that '"[t]here are people in New York that feel closer to people in London and Berlin than they do to people in Kansas and in Colorado"'; and Tharoor suggests that 'metropolises such as London [and New York] seem increasingly detached from the right-wing populist surge [represented by Brexit and Trump] in the hinterlands around them where voters rallied to calls to "take our country back" and reclaim national sovereignty'.

© The Author(s) 2017
S. Fernando, *Institutional Racism in Psychiatry and Clinical Psychology*, Contemporary Black History, DOI 10.1007/978-3-319-62728-1_9

9.1 New Era of Unashamed Racism?

Within days of his inauguration as President of the USA, Trump made it clear that the war on terrorism had moved a step further into becoming a war on Islam (Shariatmadari 2017) that meant considering all Muslims as the enemy, akin to the racist approach taken by Nazis towards Jews in the 1930s; and some of Trump's team are anti-Semitic and racist. But it is not yet clear whether Trump has the ability, sufficient support—infrastructure—or the guile to circumvent the checks and balances of the American constitution in order to pursue his policies in the face of opposition from both liberal and conservative forces inside the USA. There is a sudden realisation among liberal-minded people that the racist right wing has been boiling away almost unseen over the past ten years on both sides of the Atlantic (Economist 2016; Gray 2015; SPLC 2016) and that Trump is the political front for the so-called 'alt-right' (Caldwell 2016) which plans to manipulate the understandable anger of people in those parts of the USA that Obama stated (during the 2016 presidency campaign) 'feel left out economically and culturally' (King 2016).

Perhaps a complacency derived from having a black leader (Obama) between 2008 and 2016 blinded progressive people in the USA (and even in Europe) to the fact of *continuing* racism—which even grew stronger during the Obama years, *because* of the Obama years (see 'Obama years' in Chap. 8). After all, police shootings of African Americans in the USA occurred in great numbers *during* the Obama presidency and Obama seemed unable to do anything about it. Gary Younge (2017), editor-at-large for *The Guardian*, postulates that the election of Trump represents a reaction to Obama, to the 'past eight years [when] American liberals have gorged themselves on symbolism… [and] felt better about their country even as they fared worse in it' (p. 23). Paul Beatty, winner of the prestigious British Booker prize for his novel *The Sellout* (Beatty 2015), suggests in an interview (with Doshi 2017) that African Americans sense that extreme degrees of racism 'have always existed', and never went away. Whatever it represents, the demagoguery of Trump appears to have freed many white people to voice racist sentiments in public. In the case of the UK, similar sentiments appear to have been dormant in the imagination and have not been voiced because of political correctness (of language—see Weigel, 2016) but (as in the USA) have been released by rabble-rousing politicians, in this case from the United Kingdom Independence Party (UKIP) who, during the 'Leave' campaign for British exit from the European Union ('Brexit'), were also supported in their aims by the right wing of the Conservative party.

There was a feeling in the late 1990s, and even into the early 2000s, that most people in Europe and the USA, including people in power, were getting used to living in multiracial and multicultural societies—that some sort of accommodation had been reached among the various social and ethnic groups. And a feeling had grown that perhaps we were moving toward an era that could even be called 'post-race' ('Obama years' in Chap. 8). Suddenly this idea seems well and truly buried. The fear that is expressed openly today is that we may be on the verge of a new era of unashamed racism reminiscent of the pre-WWII years (see Chap. 3). As we head towards the third decade of the twenty-first century, over seventy years after the end of WWII, which marked a turning point in geopolitical affairs and relations between the West and the racial 'Other' (see 'Transformations after WWII' in Chap. 4), I speculate on possible directions for the future, focusing mainly on the UK, my home country for the past sixty years.

Until 9/11 the trajectory of race matters in the Western world seemed to have been moving gradually away from racism (Chap. 8). There was then a turning back, but hope was raised that this move away from racism would be resumed when Obama was elected president (in 2008) although doubts were also evident (see 'Obama years' in Chap. 8). But the events of 2016 and 2017 indicate something very different is happening. Ibram Kendi (2017), Professor of History at the University of Florida, cites outgoing President Obama's comment in his farewell speech (reproduced in The New York Times 2017) that "'for every two steps forward, it often feels like we take one step back'", and he writes:

> Mr. Obama sees in America's messiness and complexity a *single* historical force taking steps forward and backward on race. But what if there have been *two* historical forces at work: a dual and dueling history of racial progress and the simultaneous progression of racism? What if President Trump does not represent a step back, but a step forward? Americans have been well schooled in racial progress. That progress has been real over the course of history, and to deny its forward march is to deny all the successes of courageous activists who challenged slavery, and who are challenging segregation and poverty and the 45th president [Trump] today. (Kendi 2017, emphasis in original)

Many people (including myself) may have underestimated the power of racism, its *sustainability*, disregarding the significance of the racisms that were active but not easily evident (see Chap. 4) and the persistence of right-wing political movements in democratic countries of the West. As Kendi (2017) states: 'When the Obamas of the nation have broken through racial barriers, the Trumps of the nations did not give up. They organized and sometimes succeeded in putting new racial barriers in place, new discriminatory policies in our institutions. And they succeeded in developing a new round of racist ideas to justify those policies, to redirect blame for racial disparities away from their new policies and onto supposed black pathology.'

9.2 Why Racism Has Persisted

In a general sense—at a macro sociopolitical level—we could attribute the persistence of racism in Euro-America for so many years to the failure of democracy to ensure human rights in a sociopolitical context of persistent economic inequalities, where groups of people feel 'left out', excluded from the prosperity they see around them. Or, we could attribute the problem to the failure of societies, *our failure*, to confront the horrors of racism in the past with proper reckoning, facing up to the consequences of the past—something referred to in 'Transformations after WWII' in Chap. 4. Most likely, both explanations go some way towards helping us understand what has happened in the case of racism in the West. More specifically, anti-racist movements have allowed complacency (that racism was on the decline and will continue its downward path) to supervene over action, taken in by systemic manoeuvres that camouflaged racism over the years (see below).

The persistence of racism at a micro level—psychosocially—has resulted largely from the extent to which white supremacy has become embedded in the structures of systems in the West—and its export worldwide on the back of colonialism and later neocolonial economic influence ('soft power'). Further, the notion of white supremacy has infiltrated the minds of the people who have attained positions of power in the West and many parts of the Rest. But there is more to this than just the embedment of white supremacy. Racism's power of sustainability has been facilitated by its ability to change its form according to circumstances and sociopolitical context (see for example, 'American social studies' in Chap. 4 and Fernando (2014a) for discussion on how it manifests itself in the Third World in the global mental health movement). White supremacy works out through various forces implicit and explicit—of privilege, power and knowledge (discussed in 'Privilege and power' and 'White knowledge' in Chap. 7).

The advent of new types of racism in the 1960s (see 'New racisms in the UK' in Chap. 4) obscured the true nature of racism and misled anti-racist movements into thinking that they were achieving real sustainable progress. It seems that many people have been deceived by seeing people with black and brown skins in positions of authority and seeming influence—when these are essentially token appointments meant to appease racialised people, racially oppressed communities and anti-racist movements. Apart from providing role models for young people, anti-racist movements' activities have made little difference to the extent of institutional racism in society. And then there was (and is) the phenomenon of 'powerless visibility'. Stokely Carmichael and Charles Hamilton, two leaders of the Black Power movement in the USA who coined the term 'institutional racism' (see under 'The Macpherson report' in Chap. 6), also coined the term 'powerless visibility' to explain the failure of the civil rights movement in the USA to tackle institutional racism (Ture and Hamilton 1992—Stokely Carmichael having changed his name to Kwame Ture): '[W]e clearly warned that visibility did not equal power… [and]… showed integration as an insidious subterfuge for the maintenance of white supremacy…. [T]he more [black] mayors we get the more wretched becomes the condition of the masses' (p. 190). In the UK too, black and brown people are visible in many facets of British society, some in influential positions (although not yet as head of state). But apart from providing role models for young people their activities seem to make little difference in the struggle against racism in many fields, including that of mental health.

9.3 Future of the 'Psy' Disciplines

Advances in knowledge in the neurosciences are likely to have a significant impact on clinical practice in both psychiatry and clinical psychology in the future. In both disciplines, we are already seeing splits between professionals who favour looking towards greater emphasis on understanding brain function (neuroscience) (Craddock et al. 2008) and those who favour a more social-cultural approach (as opposed to a biomedical one) with an emphasis on interactions with clients (users of services) and their families (Bracken et al. 2012)—a situation I have discussed elsewhere (Fernando 2014b). As Foucault (2006) stated about clinical psychology—and the same could be said for psychiatry—it is located 'between the subject and the object' (p. 530) (see 'Sociopolitical context' in Chap. 2). So the split is essentially between the subjectivists and the objectivists.

Advances in neuroscientific knowledge that are clinically applicable are likely to drive each discipline, psychiatry and clinical psychology, at least in their clinical practices, towards a split into neuroscientific objectivity and sociocultural subjectivity. The former (neuroscientific psychiatry), hopefully dependent on diagnoses based on objective tests and the results of biological investigations (unlike the current system of biomedical psychiatry), may well be relatively free of racist influence—or at least no more racist than highly technological medical sciences. But, it would be appropriate only for the relatively few people who suffer from narrowly neuro-scientifically defined 'mental illness', with clear objective signs of physical malfunction in the brain. The bulk of the work of the mental health services will continue to concern itself with people whose problems stem from social and psychological pressures or traumas, which will continue to be given specific diagnoses or formulations based on psychological theories (see 'The scientific paradigm' in Chap. 2). There is little doubt that racism will continue to affect these services, perhaps even more so than it does in current practice. This is where anti-racist campaigns will need to focus in the continuing struggle. In other words, in spite of the likely changes in the field of mental health and the practice of psychiatry, the place of racism in clinical psychology and psychiatry will be little different to what it is today. The struggle against racism will have to continue.

9.4 Conclusions

When I began this book I had hoped to end it on a high note—suggesting confidently that the struggle against racism in the clinical disciplines of psychology and psychiatry will follow the general trend in Euro-America and continue in an upward trajectory, albeit with ups and downs. As I reached Chap. 8, seeing the exacerbation of racism around me, I had serious doubts. Since Trump and Brexit these doubts are nagging, only slightly tempered by indications that Trump and his racist colleagues in power may after all be controlled to some extent by the checks and balances of the American constitution; while the future in the UK offers little hope, since right-wing political forces seem to be gaining political power and the constitutional checks and balances to control the forces that have been released by Brexit may well be inadequate. The changed situation on both sides of the Atlantic means that the struggle against racism will be much more difficult in the future, *but is certainly not lost*. Lessons have to be learned as to why racism persisted (see the earlier section in this chapter,

'Why racism has persisted') and current political realities have to be faced, but there is no reason why the struggle should not continue. A vast literature has developed over the many years of racist hegemony to explore the numerous facets of racism and there is no need to re-explore this field yet again from an academic point of view, hoping for new conceptual tools. We have all the theories we require; the need now is to consider strategies suited to the current situation with which to resume and continue to pursue the struggle against racism.

In the UK, racialised people clearly need to reconsider the effectiveness of the ways of opposing racism that we have become used to—highlighting instances of racial discrimination and looking to legal remedies; publicising injustices in the hope of shaming people in positions of power and influence; lobbying governments; forming alliances with left-wing groups linked to demonstrations and marches; and hoping for sympathetic people in high places to support our efforts. All this has to be rethought and reappraised in the light of how and why racism has persisted for so long (see 'Why racism has persisted' earlier this chapter), and of the political events of the past twenty years, especially the advent of Trump and the post-Brexit era in Europe.

In the field of mental health, what may be needed is a diversity of approaches: focusing on thinking about the place of racism in the sociocultural dimensions of the 'psy' disciplines—and trying to bring about changes at those levels, rather than confronting racism head-on. It may be that, for a time, talking of 'race' itself will have to be lessened to concentrate on opposing the power structures of the 'psy' disciplines, whilst seeking alliances with people critical of these disciplines from other angles too. It may be necessary to focus on undermining the extent of white supremacy in the knowledge base of the disciplines (which means making these disciplines truly multicultural), and the whiteness of knowledge (see under 'White knowledge' in Chap. 7), thereby concentrating on changes within academia, which may well be more receptive to liberal ideas than the professional groups involved in practical service provision are.

I recall that in the 1960s, progressive thinkers working as psychiatrists in the asylums, frustrated at being unable to make realistic changes that would improve the asylum experience of inmates, decided that the asylum system had to be dismantled before any progress could be made towards a humane mental health system. It may be that the present system of psychiatry and clinical psychology are so bound up with racism inherited from the past *and* maintained by current social pressures (for example, pressure to keep white privilege and the power of the professional) that, as in the case of asylums in their day, the sort of psychiatry that came about following the 'medication revolution' of the 1970s (Fernando 2014a, p. 83) and the clinical psychology that reflects biologised thinking about the mind (see 'Biologisation of mind' in Chap. 2) will have to be dismantled—or perhaps paradigmatic changes have to come about in both 'psy' disciplines—before racism can be dealt with. This means that the anti-racism struggle has to collaborate with the efforts currently being made to bring about fundamental changes in psychiatry and clinical psychology.

In the 1980s, organised critique of the 'psy' disciplines from *within* the disciplines was centred, in the UK, on the Transcultural Psychiatry Society (TCPS) (see 'Transcultural psychiatry in the UK' in Chap. 6); today it is pursued by a network centred on the Critical Psychiatry Network (CPN). Unlike the case of TCPS in the 1980s and 1990s there is reluctance within CPN to grasp the nettle of race issues in public, although some individual members of the network would like that to happen. Those within the CPN who recognise the problem for what it is—racism that is deeply embedded in the machinery of current clinical practice—would like to change the 'psy' disciplines radically (see Bracken et al. 2012). However, there is also a practical need for one or more organisations run by and for racialised groups, focusing on race issues, that can lobby for change; and for non-governmental organisations to replace the black voluntary sector (BVS) which was invaluable in the 1980s and 1990 (see 'Black voluntary sector' in Chap. 6 and reference to its loss in 'Obama years' in Chap. 8). Similarly African Americans and other racialised groups in the USA need to build on movements like Black Lives Matter (Sidner and Simon 2015) (see 'Diasporic identities, nationalisms and multiculturalism' in Chap. 8) and reconsider ways of opposing racism in the mental health field taking into consideration the health systems that operate in the different states of the USA.

Historically, black people have suffered most from racism in the mental health field (see 'Race psychology' and 'Mental pathology and the construction of race-linked illnesses' in Chap. 3; 'Alleged mentality of black people' in Chap. 4; and 'Explanations for "schizophrenia" in black people' and 'Racialisation of the schizophrenia diagnosis' in Chap. 5). Today, in the mental health sector in the UK, it is being said that Muslims are the new blacks. Many psychiatric and clinical psychology colleagues (who otherwise recognise and try to minimise racism) do not seem to appreciate this or at least seem to be unwilling to voice any views on the topic. In the short term, in a context of rising racism, the 'psy' disciplines may well get drawn into active collusion in British racism's latest preoccupation, Islamophobia. Some psychiatrists already show an inclination to jump on what may soon become a research bandwagon, carrying attractive funding, that tries to reconstruct beliefs and behaviour into symptoms of illness (see 'The "psy" disciplines and Islamophobia' in Chap. 8)—just as Soviet psychiatrists once 'found' that the behaviour and beliefs of political dissidents were symptoms of schizophrenia (see 'Discrimination, diagnosis and power' in Chap. 5). I sincerely hope that the professional disciplines I have been involved in all my working life manage to keep clear of become enmeshed on the wrong side of what is likely to be a sustained demonisation of Muslims in the West. In spite of recent setbacks the likelihood is that in the UK we have come so far from racism since the end of WWII that, sooner or later, the struggle against racism in the 'psy' disciplines will resume and we will move forward again.

References

Beatty, P. (2015). *The Sellout*. London: Oneworld.
Bracken, P., Thomas, P., Timimi, S. Asen, E … and Yeomans, D. (2012). Psychiatry beyond the current paradigm. *British Journal of Psychiatry, 201*, 430–434.
Caldwell, C. (2016). What the Alt-Right really means', *The New York Times*, 2 December 2016. Retrieved on 23 January 2017 from https://www.nytimes.com/2016/12/02/opinion/sunday/what-the-alt-right-really-means.html?_r=0.
Craddock, N., Antebi, D., Attenburrow, M-J, … and Zammit, S. (2008). Wake-up call for British psychiatry. *British Journal of Psychiatry, 193*, 6–9.
Doshi, V. (2017). Paul Beatty: For Me Trump's America has always existed. Interview published in *The Observer* 22 January 2017 reproduced in *The Guardian on line*. Retrieved on 24 January 2017 from https://www.theguardian.com/books/2017/jan/22/paul-beatty-trumps-america-has-always-existed.
Fernando, S. (2014a). *Mental health worldwide. Culture, globalization and development*. New York: Palgrave Macmillan.
Fernando, S. (2014b). Concluding remarks: Future directions of psychiatry and mental health. In Moodley and Ocampo (Eds.), *Critical Psychiatry and Mental Health. Exploring the work of Suman Fernando in clinical practice* London: Routledge.
Foucault, M. (2006). *History of Madness* ed. Jean Khalfa, trans. J. Murphy and J. Khalfa (London and New York: Routledge).

Gray, R. (2015). 'How 2015 fueled the rise of the freewheeling, white nationalist alt right movement', BuzzFeedNews, 28 December 2015. Retrieved on 23 January, 2016 from https://www.buzzfeed.com/rosiegray/how-2015-fueled-the-rise-of-the-freewheeling-white-nationali?utm_term=.uxwrEWAQx#.gxwOwargz.

Kendi, I. X. (2017). Racial progress is real. But so is racist progress. *The New York Times* (New York edition) Sunday Review Opinion, 21 January 2017. Also available as 'Racial Progress, Then Racist Progress' on Page SR4 of New York Edition. on 22 January 2017 at: https://www.nytimes.com/2017/01/21/opinion/sunday/racial-progress-is-real-but-so-is-racist-progress.html?_r=0.

Khaleeli, H. (2016). Campaigners and victims reporting a rise in racist abuse since the Brexit vote. Has the hatred always been there under the surface—and will this "celebratory racism" cause lasting damage? *The Guardian supplement* 30 June 2016. pp. 6–9. Retrieved on 10 April 2017 from '"A Frenzy of hatred": how to understand Brexit racism'. https://www.theguardian.com/politics/2016/jun/29/frenzy-hatred-brexit-racism-abuse-referendum-celebratory-lasting-damage.

King, A. (2016). Obama: Trump appeals to folks who feel left out. *CNN News* 4 September 2016. Retrieved on 1 February 2017 from http://edition.cnn.com/2016/09/04/politics/obama-donald-trump-voters/.

Lakshmi, R. (2015). A lynching over beef-eating is part of a rising tide of Hindu nationalism in Modi's India. *The Washington Post*, 5 October 2015. Retrieved on 3 May 2017 at: https://www.washingtonpost.com/news/worldviews/wp/2015/08/17/how-life-in-india-has-changed-under-modi-and-why-some-muslims-arent-happy-about-it/?utm_term=.40edf20033f1.

Neslen, A. (2014). French National Front launches nationalist environmental movement. *The Guardian* 18 December 2014 Retrieved on 3 May 2017 at: https://www.theguardian.com/environment/2014/dec/18/french-national-front-launches-nationalist-environmental-movement.

Nougayrède, N. (2017). Trump's attack on human rights will be felt the world over', *The Guardian*, 23 January 2017, 23. Also available as 'Human rights now face their gravest threat—Trumpism' on 26 January 2017 at: https://www.theguardian.com/commentisfree/2017/jan/23/human-rights-threat-trumpism-white-house.

Shariatmadari, D. (2017). Trumps war on Islam. *The Guardian* g2, 31 January 2017. Also available as 'How the war on Islam became central to the Trump doctrine' on 1 February 2017 from https://www.theguardian.com/us-news/2017/jan/30/war-on-islam-central-trump-doctrine-terrorism-immigration.

Sidner, S., & Simon, M. (2015). The rise of black lives matter: Trying to break the cycle of violence and silence. *CNN*, 28 December 2015. Retrieved on 10 January 2017 from http://edition.cnn.com/2015/12/28/us/black-lives-matter-evolution/.

SPLC (Southern Poverty Law Centre). (2016). Alternative right. Retrieved on 23 January from https://www.splcenter.org/fighting-hate/extremist-files/ideology/alternative-right.

Tharoor, I. (2016). The West's major cities are a bulwark against the tide of right-wing nationalism. *The Washington Post*, 22 November. Retrieved on 24 January 2017 at: https://www.washingtonpost.com/news/worldviews/wp/2016/11/22/the-wests-major-cities-are-the-best-defense-against-the-tide-of-right-wing-nationalism/?utm_term=.4bc16ad45baf.

The Economist. (2016). Meet the IB, Europe's version of America's alt-right; From France to Austria, the "identitarian movement" gives xenophobia a youthful edge. *The Economist*, November 12. Retrieved on 23 January 2017 from http://www.economist.com/news/europe/21709986-france-austria-identitarian-movement-gives-xenophobia-youthful-edge-meet-ib.

The New York Times. (2017). President Obama's Farewell Speech. *The New York Times*, 10 January 2017. Retrieved on 23 January from https://www.nytimes.com/2017/01/10/us/politics/obama-farewell-address-speech.html?_r=0.

Ture, K., & Hamilton, C. V. (1992). *Black Power the Politics of Liberation in America new edition with new afterwords by the authors*. New York: Vintage Books a division of Random House.

Weigel, M. (2016). Political correctness: how the right invented a phantom enemy. The Guardian, 30 November 2016. Retrieved on 10 December 2016 from https://www.theguardian.com/us-news/2016/nov/30/political-correctness-how-the-right-invented-phantom-enemy-donald-trump.

Bibliography

Angell, M. (2011a, June 23). The epidemic of mental illness: Why? *The New York Review of Books*.

Balibar, E. (1991). Is there a Neo-Racism? In E. Balibar & I. Wallerstein (Eds.), *Race, nation, class. Ambiguous Identitie* (Chris Turner, Trans., pp. 17–28). London: Verso.

Balibar, É. (2015). Interview with *Relations* magazine about racism, nationalism and xenophobia in the aftermath of the Paris attacks (C. Petitjean, Trans.), *Verso Blog*, November 17, 2015. Retrieved on August 22, 2016 from http://www.versobooks.com/blogs/1559-etienne-balibar-war-racism-and-nationalism.

BBC News online. (2011). England riots: 'The whites have become black' says David Starkey. *BBC News*. August 13, 2011. Retrieved on October 10, 2016 from http://www.bbc.co.uk/news/uk-14513517.

Billig, M. (1979) *Psychology, racism and fascism*. Birmingham: A. F. and R. Publications. Retrieved on July 26, 2016 from http://www.psychology.uoguelph.ca/faculty/winston/papers/billig/billig.html.

Brittan, A., & Maynard, M. (1984). *Sexism, racism and oppression*. Oxford: Blackwell.

Collective UCL (The 'Why is my Curriculum White' collective University College London). (2015). 8 reasons the curriculum is white. *Novara Media*. Retrieved on September 27, from http://novaramedia.com/2015/03/23/8-reasons-the-curriculum-is-white/.

Collins, P. Y., Patel, V., Joestl, S. S., March, D., Insel, T. R., & Dar, A. (2011). Grand challenges in global mental health. *Nature, 475*, 27–30.

Darwin, C. (1901). *On the origin of species by means of natural selection or the preservation of favoured races in the struggle for life*. London: Ward, Lock & Co.

Das, A., & Rao, M. (2012). Universal mental health: Re-evaluating the call for global mental health. *Critical Public Health, 22*(4), 183–189.

Dodd, V. (2016). Police study links radicalisation to mental health problems. *The Guardian*, May 20, 2016. Available on January 12, at: https://www.theguardian.com/uk-news/2016/may/20/police-study-radicalisation-mental-health-problems.

DOH (Department of Health). (1998). Frank Dobson outlines third way for mental health. *Press release on Wednesday 29 July 1998. (reference number 98/311)*. Retrieved on 10 April 26 from http://www.dh.gov.uk.

Elisseeff, V. (2000). *The Silk Roads: Highways of culture and commerce*. New York: Berghahn Books.

Elkins, C. (2005). *Imperial reckoning the untold story of Britain's Gulag in Kenya*. New York: Henry Holt.

Fernando, S. (2014b). Globalization of psychiatry—A barrier to mental health development. *International Review of Psychiatry, 26*(5), 551–557.

Fernando, S., & Moodley, R. (in press). *Global psychologies: Mental health and the global south*. Basingstoke: Palgrave Macmillan.

Freely, J. (2015). *Light from the east. How science of Medieval Islam helped to shape the western world*. London: I. B. Taurus.

Gilroy, P. (2000). *Against race. Imagining political culture beyond the color line*. Cambridge, MA: Belkap Press of Harvard University.

Gray, R. (2015). How 2015 fueled the rise of the freewheeling, white nationalist alt right movement. *Buzz Feed News*, December 28, 2015. Retrieved on January 23, from https://www.buzzfeed.com/rosiegray/how-2015-fueled-the-rise-of-the-freewheeling-white-nationali?utm_term=.uxwrEWAQx#.gxwOwargz.

Hall, C. (2016). The racist ideas of slave owners are still with us today. *The Guardian*, September 27, p. 28. Retrieved on September 19, 2016 from https://www.theguardian.com/commentisfree/2016/sep/26/racist-ideas-slavery-slave-owners-hate-crime-brexit-vote.

Hall, S. (1996). The west and the rest: Discourse and power. In S. Hall, D. Held, D. Hubert & K. Thompson (Eds.), *Modernity. An Introduction to Modern Societies* (pp. 184–227). Malden MA: Blackwell.

Hamilton, P. (1996). The enlightenment and the birth of social science. In S. Hall, D. Held, D. Hubert, & K Thompson (Eds.), *Modernity. An introduction to modern societies* (pp. 19–54). Malden MA: Blackwell.

Hickling, F. W., McKenizie, K., Mullen, R., & Murray, E. (1999). A Jamaican psychiatrist evaluates diagnoses at a London psychiatric hospital. *British Journal of Psychiatry, 175*, 283–285.

Home Office. (1976). *Racial discrimination. A guide to the Race Relations Act 1976*. London: Home Office.

Home Office and the Central Office of Information. (1977). *Racial discrimination. A guide to the Race Relations Act 1976*. London: HMSO.

Kardiner, A. (1939). *The individual and his society. The psychodynamics of primitive social organisation*. New York: Columbia University Press.

Kendi, I. X. (2016). *Stamped from the beginning: The definitive history of racist ideas in America.* New York: Avalon Publishing.

Kleiner, Y. S. (2010). Minority profiling in the name of national security: Protecting minority travelers' civil liberties in the age of terrorism. *Boston College Third World Law Journal, 30*(1), 103–145. Available on October 10, 2016 at: http://lawdigitalcommons.bc.edu/twlj/vol30/iss1/5.

Koyré, A. (1954). Introduction. In E. Anscombe & P. J. Geach (Eds.), *Descartes Philosophical Writings* (pp. vii–xliv). London: Nelson University Paperbacks for Open University.

Ledwith, M. (2011). *Community development. A critical approach* (2nd ed.). Chicago: Policy Press.

Leff, J. (1974). Transcultural influences on psychiatrists' rating of verbally expressed emotion. *British Journal of Psychiatry, 125,* 336–340.

Leff, J. (1975). "Exotic" treatments and western psychiatry. *Psychological Medicine, 5,* 8–125.

Leff, J. (1981). *Psychiatry around the globe.* New York: Dekker.

Littlewood, R., & Lipsedge, M. (1982). *Aliens and alienists.* Harmondsworth: Penguin.

Martín-Baró, I. (1994). Toward a liberation psychology Ignacio Martín-Baró (A. Aron, Trans., in A. Aron and S. Corne (Eds.)), *Writings for a Liberation Psychology* (pp. 17–37). Cambridge, MA: Harvard University Press.

Martinez-Conde, S., & Macknik, S. L. (2017). The delusion of alternative facts. *Scientific American Blog Network*, January 27, 2017. Retrieved on February 10, 2017 from https://blogs.scientificamerican.com/illusion-chasers/the-delusion-of-alternative-facts/?print=true.

McManus, I. C. (1998). Factors effecting likelihood of applicants being offered a place in medical schools in the United Kingdom in 1996 and 1997: Retrospective study. *British Medical Journal, 317,* 16–1111.

MHAC (Mental Health Act Commission). (1989). *Third Biennial Report 1987–1989.* London: HMSO.

Modood, T. (1997). Employment. In T. Modood, R. Berthoud, J. Lakey, P. Smith, V. Satnam, & S. Beishon (Eds.), *Ethnic minorities in Britain. Diversity and disadvantage* (pp. 83–149). London: Policy Studies Institute.

Morrison, T. (1993). *Playing in the dark. Whiteness and the literary imagination.* London: Pan Macmillan.

Nazroo, J. Y. (2015). Ethnic inequalities in severe mental disorders: Where is the harm? *Social Psychiatry and Psychiatric Epidemiology, 50,* 67–1065.

Patterson, O. (2011). The post-black condition. *The New York Times,* September 22, 2011. Retrieved on January 22, 2017 from https://www.nytimes.com/topic/subject/race-and-ethnicity?action=click&contentCollection=Sunday%20Book%20Review&module=RelatedCoverage&region=EndOfArticle&pgtype=article.

Performance and Innovation Unit. (2001). *Scoping note. Improving labour market achievements for ethnic minorities in British society*. Retrieved on June 1, 2012 from http://www.cabinet-office.gov.uk/innovation/test/scope.html.

Perneger, T. V. (1998). What's wrong with Bonferroni adjustments. *British Medical Journal, 316*(7139), 1236–1238. Retrieved on December 10, 2016 from https://www.ncbi.nlm.nih.gov/pmc/articles/PMC1112991/.

Phillips, J. (2010). *Holy warriors: A modern history of the crusades*. London: Vintage.

Quarashi, F. (2016). Prevent gives people permission to hate Muslims—it has no place in schools. *The Guardian*, April 4, 2017. Retrieved on January 12, 2016 from https://www.theguardian.com/commentisfree/2016/apr/04/prevent-hate-muslims-schools-terrorism-teachers-reject.

Quassem, T., Bebbington, P., Spiers, N., McManus, S., Jenkins, R., & Dein, S. (2015). Prevalence of psychosis in black and ethnic minorities in three national surveys. *Social Psychiatry and Psychiatric Epidemiology, 50*, 64–1057.

Qureshi, A. (2016). *The 'Science' of Pre-Crime. The Secret 'Radicalisation' Study Underpinning Prevent*. London: Cage. Retrieved on April 1, 2017 from http://cage.ngo/publication/the-science-of-pre-crime/.

Ross, A. (2016). Academics criticise anti-radicalisation strategy in open letter. *The Guardian*, Thursday September 29, 2016, Retrieved on April 4, 2017 from https://www.theguardian.com/uk-news/2016/sep/29/academics-criticise-prevent-anti-radicalisation-strategy-open-letter.

Ryle, G. (1990). *The Concept of Mind*. London: Penguin Books (First published by Hitchingson, New York, 1949).

Scott, P. (1996). *The jewel in the crown*. London: Heinemann.

Select Committee on Race Relations and Immigration. (1977). *The West Indian Community*. London: Her Majesty's Stationery Office.

Silverstein, P. A. (2005). Immigrant racialization and the new savage slot: Race migration and immigration in the New Europe. *Annual Review of Anthropology, 34*, 363–384. Retrieved on February 10, 2017 from http://www.annualreviews.org/doi/full/10.1146/annurev.anthro.34.081804.120338.

The Telegraph. (2007) Enoch Powell's "River of Blood" speech. *The Telegraph*, November 06, 27. Retrieved on February 10, from http://www.telegraph.co.uk/comment/3643823/Enoch-Powells-Rivers-of-Blood-speech.html.

Thubron, C. (2007). *Shadow of the Silk Road*. Vancouver: Vintage Books.

Wintour, P. (2014). Blunket urged to resist migrant crackdown. *The Guardian*, 23 February 24. Retrieved on December 10, 2016 from https://www.theguardian.com/politics/24/feb/23/eu.thinktanks.

World Health Organisation. (1973). *Report of the International pilot study of Schizophrenia* (Vol. 1). Geneva: WHO.

Zureik, E. (2015). *Israel's Colonial project in Palestine: Brutal pursuit*. London: Routledge.

Author Index

A
Aitkenhead, D., 97
Alexander, M., 156
Allen, T.W., 41
Anderson, B., 154
Angell, M., 34
Anon, 78
Antebi, D., 185
Anwar, M., 139
Arendt, H., 60
Armstrong, K., 171
Asen, E., 185, 188
Attenburrow, M-J., 185

B
Babcock, J.W., 52
Ballantyne, A., 104
Banton, M., 11
Barker, P., 163
Barzun, J., 12, 21
Bauman, Z., 154
Bayer, R., 51
Bean, R.B., 19
Beatty, P., 182
Bebbington, P., 82, 83
Beresford, P., 163
Berman, G., 137
Bernal, M., 16, 44
Beydoun, K.A., 168

Bhabha, Homi K., 85, 154
Bhattacharyya, G., 146
Bhui, K., 172
Billig, M., 44, 47, 77
Blackman, P., 163
Blackstock, C., 160
Bleuler, E., 106
Bloch, S., 92
Bloombaum, M., 95
Bonilla-Silva, E., 19
Bourne, J., 71
Bracken, P., 185, 188
Bright, T., 21
Brightwell, R., 104
Brignell, V., 47
Brittan, A., 172
Brodkin, K., 42, 97
Burton, R., 21
Bush, George, 168

C
Caldwell, C., 182
Campling, P., 95
Cannon, M., 104
Cantor-Graae, E., 105
Carew, J., 13
Carmichael, S., 64, 68, 73, 111, 185
Carothers, J.C., 80–82
Cartwright, S.A., 51

Christie, Y., 163
Clarke, J., 115
Clear, P., 163
Clifford, T., 32
Coffey, V.P., 104
Cohen, P., 161
Cole, S., 169
Collins, P.Y., 184
Cooke, A., 163
Coon, C.S., 65
Cooper, S.J., 83, 104
Cope, R., 94, 105
Cotter, D., 104
Cotterell, A., 39
Cox, J., 113
Craddock, N., 185
Crepaz-Keay, D., 163
Critcher, C., 115
Croft-Jeffreys, C., 100
Crow, Jim, 1, 12, 15, 62, 106, 157
Crow, T.J., 104
Cuffal, B., 94

D
Dabashi, H., 156
Dalal, F., 53, 54
Dallos, R., 25
Dalrymple, W., 39
Dar, A., 137
Darwin, C., 44, 47, 51
David, A., 100
Davidson, B., 40, 43
Davies, P., 30
Davies, S., 100
Dazzan, P., 104, 162
Dean, C., 171
Dein, S., 83
de Martino, R., 27
Diesenhaus, H., 105

Disraeli, B., 17, 39
Dodd, V., 112
Dols, M.W., 22, 32, 170
Donald, A., 29
Done, D.J., 104
Doshi, V., 182
Down, J.L.M., 50
du Guy, P., 154
Durrheim, K., 72

E
Eagles, J.M., 104
Easterly, W., 40, 129
Elkins, C., 60
El Magd, N.A., 148
Erichsen, C.W., 41
Erickson, E.P., 74
Essed, P., 72, 85
Evans, R., 112
Everitt, B., 172
Eysenck, H.J., 77
Eze, E.C., 18, 19

F
Fanon, F., 3, 75, 81, 96, 143
Faulkner, A., 165
Fearon, P., 104, 162
Ferns, P., 163
Finocchiaro, M.A., 23
Fitzpatrick, P., 96
Foote, R.F., 46
Ford, P., 8
Foucault, M., 21, 27–30, 32, 92, 170, 185
Frank, J.D., 95
Fromm, E., 27
Fryer, P., 19
Fung, W.A., 104, 162

G

Gabriel, J., 146
Galton, F., 26, 45, 47
Gama, Vasco de, 15
Garner, S., 68, 73, 112
Geary, D., 68, 77
Gergen, K.J., 99
Gibson, C., 59
Gillis, L.S., 93
Gilman, S., 45
Gilroy, P., 85
Goldberg, D.T., 17, 72
Gordon, L., 11
Gradwell, S., 163
Graham, T.F., 22, 32
Gray, R., 182
Gribbin, J., 30

H

Hacking, I., 33
Haeckel, E., 44
Haig-Brown, C., 31
Hall, C., 45
Hall, G.S., 51
Hall, R., 16
Hall, S., 11, 71, 72, 74, 84, 115, 154
Hamilton, C.V., 64, 73, 111, 185
Harari, E., 94, 98
Harrison, G., 94, 100, 104
Hasan, M., 167
Hattem, J., 95
Hegel, G.W.F., 43
Hepler, N., 94
Hepple, B.A., 1, 75
Hero, R.E., 156
Hochschild, A., 41
Hollady, J., 94
Holloway, J., 104, 162
Hook, D., 72
Howe, D., 70
Howitt, D., 147

Huntingdon, S.P., 168
Hurry, J., 83
Hutchinson, G., 104, 162
Hutton, W., 158

I

Ignatief, N., 42, 97
Imlah, N., 105
Ineichen, B., 94
Ingleby, D., 24
Inyama, C., 141

J

James, Q.C., 95
Jarvis, E., 78
Jefferson, T., 85, 115
Jensen, A.R., 76
Johnstone, L., 25
Jolliff, T., 138
Jones, E., 172
Jones, P.B., 104, 162
Jordon, W.D., 14
Julian, R.N., 74
Jung, C.G., 52, 53, 161
Jung, K., 59

K

Kabbani, R., 45
Kalathil, J., 164, 165
Kamen, H., 12
Kamin, L.J., 77
Kapuściński, R., 11, 33
Kardiner, A., 65
Karenga, M., 31
Karr, A., 172
Karr, J.-B. A., 154
Keating, F., 120
Keller, R.C., 81, 82
Kendi, I.X., 183, 184

Keval, H., 166
Khaleeli, H., 167, 181
Khalfa, J., 28
King, A., 182
King, D.J., 104
Kirkbride, J.B., 105
Kleinman, A., 99, 100
Kline, R., 138
Knox, H.G., 48
Kollie, A., 163
Koyré, A., 23
Kraepelin, E., 29, 50, 51, 79, 106
Kuhn, T.S., 24
Kuller, L.H., 100

L

Lakshmi, R., 181
Larkin, C., 104
Lawrence, E., 69
Lawrence, Stephen, 111
Lawson, W.B., 94
Lears, J., 168
Ledwith, M., 122
Leese, M., 100
Leff, J., 83
Levy, M.E., 156
Lewin, M., 162
Lewis, A., 78
Lewis, G., 100
Lichtblau, E., 159
Lindsey, K.P., 94
Linnaeus, C., 19
Lloyd, M., 171
Lloyd, T., 104
Lombroso, C., 106
Loring, M., 94
Losurdo, D., 18
Lowery, W., 156

M

Mallett, R., 104
Mandela, Nelson, 60
Mansoor, S., 62
Martín-Baró, I., 24
Martin, C., 163
Marx, K., 40
Maudsley, H., 78
Maynard, M., 96
McDougal, W., 48
McGovern, D., 94, 105
McKenizie, K., 101
McManus, S., 83
McQueen, D.V., 22
Menzies, G., 13
Metta, J., 159
Metzl, J., 94, 106
Miles, R., 62
Mishra, P., 86, 153
Modood, T., 154
Moodley, R., 115
Moon, C., 60
Moore, P., 159
Moorhouse, G., 39, 40
Morel, B.-A., 106
Morgan, C., 104
Morgan, H.G, 94
Morgan, K., 104
Morrison, T., 18, 19
Moynihan, D., 66
Murphy, G., 21
Murray, E., 101
Murray, R.M., 104

N

Nazroo, J.Y., 83
Neslen, A., 181
Nougayrède, N., 181

O

Obama, Barack, 8, 155, 156, 158, 182, 183
O'Callaghan, E., 104
Ocampo, M., 115
Olusoga, D., 69, 70

Omi, M., 2, 71, 72
Outram, D., 23
Ovesey, L., 65

P
Pajaczowska, C., 160
Pakenham, T., 40
Palley, N., 95
Panikkar, K.M., 39
Parkman, S., 100
Patel, V., 184
Patterson, J., 42
Patterson, O., 183
Paul, G.L., 94
Pearson, K., 47
Perneger, T.V., 29, 46, 101
Phelan, M., 100
Phillips, J., 167
Pick, D., 26, 106
Pieterse, J.N., 154
Pilkington, E., 62
Pinderhughes, C., 93
Porter, R., 21, 22, 169
Powell, B., 62, 94, 95, 161
Prince, R., 82
Pritchard, J.C., 78

Q
Quarashi, F., 42, 138, 162, 163, 167, 182
Quassem, T., 83
Qureshi, A., 106

R
Radcliff, B., 156
RawOrg, 123
Reddaway, P., 92
Richards, G., 46–49, 77
Riggs, D.W., 72

Ringler, S., 155
Roberts, B., 85, 115
Robinson, F., 40
Rodney, W., 41
Rosenthal, D., 95
Rüdin, 50
Ryle, G., 23

S
Saathoff, G., 60
Sabshin, M., 105
Sagemen, M., 171
Said, E., 169
Sashidharan, S.P., 79, 113, 140, 142
Schwartz, K., 168
Scott, P., 39
Scull, A., 169
Searle, C., 13
Selten, J.-P., 104
Seymour, R., 158, 159
Sham, P.C., 104
Shorter, E., 28
Sidner, S., 156, 188
Sillen, S., 49, 78
Silva, M.J., 172
Silverman, J., 70
Silverstein, P.A., 19, 45, 49, 65, 79
Simon, M., 156, 188
Singh, A., 169, 171
Skultans, V., 51
Small S., 146
Smith, G., 21
Smith, J., 94
Smith A., 19
Spiers, N., 83
Spurlock, J., 93
Stanley, J., 163
Stannard, D.E., 13
Starkey, David, 62, 161
Stone, A., 93
Stonequist, E.V., 106

Stott, D.H., 77
Straw, Jack, 160
Summerfield, D., 169
Suzuki, D.T., 27

T
Takei, N., 104
Tarrant, J., 104
Taylor, R., 138
Tennant, C., 83
Terman, L.M., 49
Tharoor, Ishan, 181
Tharoor, Shashi, 16, 39
Thomas, A., 49, 78
Thomas, P., 163
Thompson, S., 163
Thornicroft, G., 100
Tilby, A., 104
Timimi, S., 163, 185, 188
Topciu, R.A., 172
Torrey, E.F., 79
Touré, 155
Toynbee, P., 159
Trevor-Roper, H., 43
Trump, Donald, 8, 153, 160, 173, 181, 182, 184, 186, 187
Tuke, D.H., 46, 78
Ture, K., 185

V
Vernon, P., 163
Vige, M., 115

W
Wallcraft, J., 164
Walwin, J., 14
Warfa, N., 172
Warner, R., 79
Watson, P., 77, 118
Weigel, M., 1, 136, 160
Weinberg, M.D., 93
Weindling, P., 50
Wellman, D., 73
Wilkerson, R., 105
Willinsky, J., 39, 43, 148
Winant, H., 2, 71, 72, 96, 98
Wing, J.K., 104
Wintour, P., 158
Woods, A., 167
Woodward, C.V., 15
Worsley, P., 43

Y
Yamamoto, J., 95
Yancey, W.L., 74
Yerkes, R.M., 49
Young, L., 45
Young, S.S., 173

Z
Zakaria, A., 62
Zammit, S., 185
Zureik, E., 5, 15

Subject Index

A

Abolition of slavery, 15, 42
Abyssinia, 40
Adolescence, 51
Aesop study, 104, 162, 165
Africa, 7, 15, 16, 18, 31, 39–41, 43, 45, 50, 53, 60, 62, 78, 81, 82
African, 6, 13–16, 28, 29, 31, 41–43, 45, 51–53, 63, 65, 66, 68, 74, 78, 80–83, 117, 124, 154, 162
African Americans, 52, 65–67, 83, 94, 118, 155, 156, 182
African-Caribbean, 68–70, 83, 105, 122, 124, 162
Africanism, 19
African mind, 80, 81
African National Congress (ANC), 117
Age of anger, 86, 153
Age of reason, 21
Algeria, 60, 62, 82
Alienists, 5, 28, 29
American blacks, 46, 79, 80, 83, 106
American constitution, 182, 186
American holocaust, 13
American Psychiatric Association (APA), 51, 93
American social studies, 77, 118, 162, 184
American whites, 46
Anti-semitism, 2, 13, 60, 155
Apartheid, 7, 60, 72, 92, 117, 172
Aryans, 44
Asia, 12, 15, 16, 18, 39–42, 45, 50, 52, 59, 60, 62, 82, 159
Asian, 7, 28, 31, 43, 69, 74, 79, 85, 95, 96, 100, 103, 118, 137, 139, 141, 144, 155, 160
Association of Medical Officers of Asylums and Hospitals, 29
Atlantic slave trade, 39, 42

B

BBC News, 104, 160, 161
Bekterev, 92
Belgian Congo, 41
Big, black and dangerous stereotype, 105
Biologization of mind, 23, 33, 188
Black and Minority Ethnic (BME), 2, 116, 138, 157
Black athena, 44
Black pathology, 184
Black power, 145, 185
Black professionals, 102, 118, 140
Black schizophrenia, 104
Black service users, 102, 165
Black skin, white masks, 3, 143
Black Voluntary Sector (BVS), 119, 188

BME, 2, 85, 96, 113, 115, 118, 119, 121–123, 126, 127, 140–145, 154, 164, 166
BME professionals, 115, 141, 144
Brexit, 8, 9, 160, 171, 181, 186
British-Chinese Asian, 69
British Eugenic Society, 65
British Psychological Society (BPS), 78, 141, 163
Brixton, 69, 70, 126
Brixton riots, 70
Burnley Task Force, 159

C

Cambodia, 62
Cambridge University, 139
Cantle Report, 159
Care Quality Commission (CQC), 130
Cartesian doctrine, 23
Category fallacy, 99
Catholic Church, 21, 23, 25
Centre for Contemporary Studies, 115
Channel Duty Guidance, 170
Chapeltown, 69
Chemical imbalance, 33, 34
Chicago, 63
China, 16, 17, 43–46
China opium, 17
Chinese, 16, 17, 32, 44, 49, 68, 98, 155
Civil rights movement, 6, 15, 59, 62, 63, 106, 185
Clinical psychology, 2, 5, 7–9, 22, 28–30, 33, 46, 78, 85, 86, 94, 113, 114, 173, 185–188
Collective UCL, 148
Colonialism, 1, 5, 8, 11, 13, 15, 16, 18, 29, 39–42, 51, 60, 61, 75, 85, 96, 97, 128, 184
Colonisation, 16, 31, 40–42, 44, 145
Colour-bar, 1, 75

Columbus, 13, 31
Commission for Racial Equality (CRE), 119, 135
Communism, 50
The Counted, 156
Counter-terrorism, 170
Criminal justice, 8, 15, 138
Cultural research, 82

D

Darwinism, 26, 47
David bennett inquiry, 123
Definition of race, 72, 98
Definitions of racism and race, 4, 76, 85, 96, 98, 111, 155
Degeneration, 106
Delivering race equality, 123, 142
Department of Education (DOE), 153
Department of Health (DOH), 75, 102, 119, 140, 145, 157
Diagnosis power, 7, 91, 117, 136, 147, 189
Diasporic identities, 153, 181, 188
Diverse minds, 115
Division of Clinical Psychology (DCP), 30, 130, 163
Dobson, 140
Drapetomania, 52

E

East India company, 39, 40
Economist, The, 61
ECT, 81
Editorial, 159
Egypt, 44
Egyptians, 44, 46
Employment industrial tribunal, 139
Enlightenment, 4, 6, 17, 18, 21, 22, 24, 105
Enslaved Africans, 14

Epidemiological studies, 100
Equality Act, 135
Equality and Human Rights Commission (EHRC), 135
Eskimos, 46
Ethiopian, 51
Ethnicity, 74, 91, 100, 103, 166
Ethnicity and diagnosis, 137
Ethnicity and 'stop and search, 137
Ethnocide, 13
Eugenic studies, 49
European values, 4, 17, 84
Everyday racism, 72, 85
Exploitation, 4, 8, 16, 18, 40, 60, 75, 121
Exploration, 16, 170

F

Faculty of race and culture, 131, 166
French revolution, 18

G

Gallileo, 23
Genocide, 13, 16, 31, 41
German Psychiatric Research Institute, 50
Glasgow, 70
Greek medicine, 22
Group minds, 48

H

Haiti, 18
Harambee, 120
Hebrew culture, 32
HM Government, 169, 170
Home Department, 73, 111
Home office, 160
Homosexuality, 51
Hottentot, 45

House of Commons, 40
Human rights, 61, 64, 154, 158, 169, 181, 184

I

ICERD, 64
Immigration, 47, 49, 62, 82, 91, 104, 158–160, 171
Independent Review Team, 159
India, 15, 16, 39, 40, 43, 62, 168, 181
Indigenous American, 14
Indonesia, 169
Industrial tribunal, 139
Inside outside, 122, 123, 142, 158
Instincts, 48
Institute of Psychiatry (IOP), 50, 69, 104, 144, 162, 165
Institutional racism, 5, 7, 63, 64, 73, 75, 84, 102, 103, 112–114, 116, 123, 124, 126–128, 130, 131, 138, 143, 160, 185
Intelligence tests (IQ), 49
International convention on the elimination of all forms of racial discrimination (ICERD), 61, 135
International Pilot Study of Schizophrenia (IPSS), 83
Ipamo project, 119, 126
IQ tests, 76
Irish, 42, 83, 97
Irish Republican Army (IRA), 97
Islamic 'psychiatry, 32
Islamic Empire, 40
Islamophobia, 157, 167–169

J

Jamaica, 101
Japan, 69
Java, 168
Jesuits, 45

Jews, 2, 13, 42, 97, 167, 170, 182
Jihad, 153, 168

K
Kenya, 60, 80, 81
Ku Klux Klan, 15

L
Leading Article, 93
Leningrad, 92
Levant, 40
Liverpool, 70
London, 2, 7, 45, 70, 97, 127, 139, 161, 162, 181
London Jewish Hospital, 2
London Metropolitan Police, 111, 112, 137
Lynfield Mount Hospital, 114

M
Macpherson report, 113, 160
Madness, 22, 26–29, 31–33, 78, 155
Maghreb, 40, 81, 82
Malay, 51
Mapping Police Violence, 156
Māristāns (Islamic hospitals), 32
Mark of oppression, 65–68
Maudsley hospital, 124, 125
Maximum security, 102
Medication revolution, 27, 33, 188
Melting pot, 47
Mental Health Act, 94, 121, 141, 142
Mental Health Act Commission (MHAC), 7, 119, 145
Mental Health Network (BMENW), 141
Mentality of black people, 28, 52, 188
Metropolitan police, 68, 73, 112, 161
Middle Ages, 12, 40, 44, 168

Middle east, 12
Middle passage, 14
Mohammedan, 46
Mongolism, 51
Moors, 11–13, 168
MOST project, 125
Moynihan report, 66, 68, 77, 162
Multiculturalism, 7, 120, 153
Muslims, 4, 13, 42, 82, 158, 159, 167–170, 172, 188

N
Nafsiyat, 120
National Archives, The, 138
National BME Mental Health Network (BMENW), 141
National Health Service (NHS), 91, 114, 138, 170
National Institute for Mental Health in England (NIMHE), 158
Nationalisms, 153, 181, 188
Nativism, 40, 41, 44, 45, 78, 81, 101
Nazi, 50, 77
Negro, 14, 19, 20, 48, 53, 65
Negro education, 46, 47
Neuroscientific knowledge, 186
New Cross, South London, 70
New racisms, 6, 64, 66, 84, 97, 161, 185
Newtonian physics, 23, 25
New York, 4, 153, 181
New York Times, The, 183
Nile centre, 120
Nine-eleven (9/11), 153
Noble Savage, 78, 79
Norfolk, Suffolk and Cambridgeshire Strategic Health Authority, 123
North Africa, 12, 17, 43, 81
Notion of 'race', 1, 2, 5, 12, 42, 74, 95, 96
Nuremburg courts, 50

O

Obama years, 5, 18, 42, 124, 156, 182, 183, 188
Oldham Independent Panel, 159
Operation Swamp, 70
The 'other', 11, 41, 165
Orient, 45, 169
Orientalism, 46, 168
Orientals, 43

P

Political correctness, 1, 6, 64, 84, 160, 173, 182
Porot, 81
Positivism, 24, 170
Power of racism, 41, 81, 145, 183
Pre-Columbian, 6, 31
Pre-Columbian America, 31
Prison population, 138, 171
Privilege, 3, 8, 71, 146, 148, 157, 166, 167, 184, 188
Psychiatric research, 7, 82, 98, 100, 103
Psychiatrists Against Apartheid (PAA), 117
Psychologizing of the self, 21
Psychology report, 78, 166
Psy disciplines, 173

R

Race-linked illnesses, 29, 50, 80, 161, 188
Race psychology, 26, 47, 49, 50, 65, 173
Race Relations Act, 1, 75, 77, 113, 119, 129, 135
Race Relations Act (1968), 75, 77
Race Relations Act (1976), 1, 63

Race-slavery, 1, 11, 12, 16, 29, 39, 145, 153, 168, 170
Race thinking, 5, 12, 29, 84
Racial 'Other', 6, 61, 97, 154, 183
Racial Awareness Training (RAT), 70
Racial discrimination, 59, 61, 63, 65, 85, 116, 121, 131, 135, 187
Racial infection, 52, 53, 161
Racialisation, 1, 96, 97, 105, 113, 145, 154, 157, 161, 188
Racialised groups, 1, 97, 105, 113, 136, 143, 147, 153–155, 188
Racism, 1–9, 13, 29, 41, 47, 184, 186–189
Racist IQ movement, 6, 49, 77, 137
Radicalisation, 157, 170, 172
Rationality, 17, 23, 24
Renaissance, 21, 44
Research, 2, 7, 19, 45, 49, 50, 84, 98, 100–103, 125, 139, 162, 168
Rivers of Blood speech, 62, 161
Rockefeller Foundation, 50
Roman North Africa, 43
Romany communities, 4
Royal College of Psychiatrists (RCP), 7, 29, 30, 93, 113, 144, 171
Runnymede Trust, 136
Russia, 169

S

Scarman, 70, 71
Schizophrenia, 7, 27, 50, 54, 79, 92, 94, 97, 99, 105, 165, 166, 188
Schizophrenia diagnosis, 106
Schizophrenia in black people, 101, 104, 163
School exclusions, 137
Scientific paradigm, 22, 24, 25, 30, 186

SUBJECT INDEX

Scientific racism, 19, 20, 76, 77
Scientific validity, 25
Scramble for Africa, 40
Second World War, 4, 59
Second World War (WWII), 4, 27, 59, 60, 62, 77, 84, 153, 167, 189
Sectioning, 121, 141
Select Committee on Race Relations and Immigration (1977), 69
Semites, 44
Serbsky Institute, 92
Sinologists, 46
Slavery, 5, 15, 18, 39, 40, 42, 65, 85, 96, 147, 172
Slave traffickers, 41
Social Darwinism, 26, 47
Sociology, 6, 45, 65
Somalia, 169
South Africa, 7, 60, 72, 92, 117, 172
Southern Poverty Law Centre (SPLC), 182
South-West Africa, 41
Soviet Union, 7, 60, 92, 172
Spain, 11–13
Spanish Inquisition, 12, 167, 170
Special hospitals, 101
Special Hospitals Service Authority (SHSA), 105, 122
Spirituality, 21, 25, 27, 31, 98, 126
The Stationery Office (TSO), 135, 170
Stereotypes, 19, 78, 81, 105, 122, 136, 147, 163, 169
Stop Watch, 138
Struggle against racism, 7, 9, 114, 117, 153, 160, 163, 185–187, 189
Submissiveness, 48, 52

T
Tangle of pathology, 66, 67, 69
Telegraph, The, 62
Theories of black racial inferiority, 66
Third world, 6, 43, 184
13th Amendment to the U.S. Constitution Abolition of Slavery, 15, 42
Tibetan psychiatry, 32
Tory government, 62, 70, 111, 119
Toxteth, 69
Traditional cultures, 31
Transcultural psychiatry, 78, 79, 114–116, 119
Transcultural Psychiatry Society (TCPS), 97, 104, 188
Tribalism, 41
Turkish hybridity, 154

U
UN, 8, 61, 135, 136
Unashamed racism, 5, 9, 76, 183
United Nations (UN), 6, 60, 61
United Nations Educational, Scientific and Cultural Organisation (UNESCO), 61, 81
University College London (UCL), 47, 148
Unwitting prejudice, 112
USA, 1, 2, 6, 7, 12, 14, 15, 26, 42, 46, 48–52, 59, 62, 63, 65, 66, 69, 73, 78, 79, 85, 94, 97, 106, 117, 145, 146, 156, 158, 160, 182, 183, 188

V
Vietnam, 62
Vulgar racism, 75

W
War against terrorism, 4
Western culture, 3, 5, 8, 9, 12, 20, 21, 31, 49, 51, 78, 81, 99, 147, 173
White-dominated systems, 8, 143

White House Archives, 168
White knowledge, 8, 93, 98, 147, 148, 166, 167, 187
White Man's Burden, 40
Whiteness, 8, 41, 44, 97, 106, 145, 146, 148, 166, 187
White privilege, 71, 146, 166, 187

White supremacy, 6, 8, 22, 143, 146, 167, 184, 187
World Health Organization (WHO), 61, 80

X
Xenophobia, 160, 181